Sexual Difference between Psychoanalysis and Vitalism

Throughout the twentieth century, psychoanalysis and feminism were the practico-intellectual fields most systematic and subversive in demonstrating that humanity is sexually fissured. More recently, further advances in the philosophy of difference and renewed emphases on embodiment, materiality and life offer possibilities for attending to dimensions of gender and sexuality that were previously underdeveloped. This collection examines these possibilities insofar as they can either deepen or displace the traditional centrality of psychoanalysis in matters sexual.

The authors come from a wide range of backgrounds and defend their approaches to the problem of sexual difference in a variety of idioms, drawing on key thinkers such as Lacan, Irigaray, Deleuze, Foucault and Badiou. It is rare to come across these thinkers together, but sex is too crucial a site for critical thought not to mobilize every conceptual power available.

This book was originally published as a special issue of the journal *Angelaki: Journal of the Theoretical Humanities.*

Arun Saldanha is Associate Professor of Geography at the University of Minnesota, Minneapolis, USA. He is the author of *Psychedelic White: Goa Trance and the Viscosity of Race* (2007) and co-editor of *Deleuze and Race* (2013) and *Geographies of Race and Food: Fields, Bodies, Markets* (2013).

Hoon Song is Associate Professor of Anthropology at the University of Minnesota, Minneapolis, USA. He is the author of *Pigeon Trouble: Bestiary Biopolitics in a Deindustrialized America* (2010).

Sexual Difference between Psychoanalysis and Vitalism

Edited by
Arun Saldanha and Hoon Song

Routledge
Taylor & Francis Group
LONDON AND NEW YORK

First published 2014
by Routledge
2 Park Square, Milton Park, Abingdon, Oxon, OX14 4RN

Simultaneously published in the USA and Canada
by Routledge
711 Third Avenue, New York, NY 10017

Routledge is an imprint of the Taylor & Francis Group, an informa business

British Library Cataloguing in Publication Data
A catalogue record for this book is available from the British Library

ISBN13: 978-0-415-83686-9

Typeset in Minion
by Taylor & Francis Books

Publisher's Note
The publisher accepts responsibility for any inconsistencies that may have arisen during the conversion of this book from journal articles to book chapters, namely the possible inclusion of journal terminology.

Contents

Citation Information vii
Notes on Contributors ix

1. One, Two, Many: what is sexual difference now?
Arun Saldanha 1

2. The Sexual Compact
Joan Copjec 30

3. Being or Sex, and Differences
A. Kiarina Kordela 48

4. The Nature of Sexual Difference: Irigaray and Darwin
Elizabeth Grosz 67

5. Sexual Difference beyond Life and Death: or, feminism and the biopolitical turn
Cesare Casarino 92

6. The Body, Sexuality, and Sexual Difference
Charles Shepherdson 101

7. Two is Infinite, Gender is Post-Social in Papua New Guinea
Hoon Song 118

8. Against Yin-Yang: the dao of feminist universalism
Arun Saldanha 140

9. Afterword: complexity, materialism, difference
Rosi Braidotti 164

Index 172

Citation Information

The chapters in this book were originally published in the *Angelaki: Journal of the Theoretical Humanities*, volume 17, issue 2 (June 2012). When citing this material, please use the original page numbering for each article, as follows:

Chapter 1
One, Two, Many: what is sexual difference now?
Arun Saldanha
Angelaki: Journal of the Theoretical Humanities, volume 17, issue 2 (June 2012)
pp. 1–30

Chapter 2
The Sexual Compact
Joan Copjec
Angelaki: Journal of the Theoretical Humanities, volume 17, issue 2 (June 2012)
pp. 31–48

Chapter 3
Being or Sex, and Differences
A. Kiarina Kordela
Angelaki: Journal of the Theoretical Humanities, volume 17, issue 2 (June 2012)
pp. 49–68

Chapter 4
The Nature of Sexual Difference: Irigaray and Darwin
Elizabeth Grosz
Angelaki: Journal of the Theoretical Humanities, volume 17, issue 2 (June 2012)
pp. 69–94

Chapter 5
Sexual Difference beyond Life and Death: or, feminism and the biopolitical turn
Cesare Casarino
Angelaki: Journal of the Theoretical Humanities, volume 17, issue 2 (June 2012)
pp. 95–104

Chapter 6
The Body, Sexuality, and Sexual Difference

Charles Shepherdson
Angelaki: Journal of the Theoretical Humanities, volume 17, issue 2 (June 2012)
pp. 105–122

Chapter 7
Two is Infinite, Gender is Post-Social in Papua New Guinea
Hoon Song
Angelaki: Journal of the Theoretical Humanities, volume 17, issue 2 (June 2012)
pp. 123–144

Chapter 8
Against Yin-Yang: the dao of feminist universalism
Arun Saldanha
Angelaki: Journal of the Theoretical Humanities, volume 17, issue 2 (June 2012)
pp. 145–168

Chapter 9
Afterword: complexity, materialism, difference
Rosi Braidotti
Angelaki: Journal of the Theoretical Humanities, volume 17, issue 2 (June 2012)
pp. 169–176

Please direct any queries you may have about the citations to
clsuk.permissions@cengage.com

Notes on Contributors

Rosi Braidotti is Distinguished University Professor and founding Director of the Centre for the Humanities at Utrecht University, the Netherlands. Her books include *Patterns of Dissonance* (1991); *Nomadic Subjects* (1994, 2011); *Metamorphoses* (2002); *Transpositions* (2006); *La Philosophie, là où on ne l'attend pas* (2009); and *Nomadic Theory: The Portable Rosi Braidotti* (2011). She has received multiple prestigious awards for her work. Since 2009 she has been a board member of the Consortium of Humanities Centers and Institutes (CHCI) at Duke University, USA.

Cesare Casarino is Professor of Cultural Studies and Comparative Literature at the University of Minnesota, Minneapolis, USA. He is the author of *Modernity at Sea: Melville, Marx, Conrad in Crisis* (2002); co-author (with Antonio Negri) of *In Praise of the Common: A Conversation on Philosophy and Politics* (2008); co-editor (with Saree Makdisi and Rebecca Karl) of *Marxism beyond Marxism* (Routledge, 1996); as well as author of essays on literature, cinema and philosophy. His most recent publications include "Marx before Spinoza (Notes toward an Investigation)" in *Spinoza Now* (2011) and "Three Theses on the Life-Image (Deleuze, Cinema, Biopolitics)" in *Releasing the Image: From Literature to New Media* (2011). At present he is at work on book manuscripts on Pasolini, on Marx and Spinoza, as well as on Deleuze's theory of the cinema.

Joan Copjec is SUNY and UB Distinguished Professor of English, Comparative Literature and Media Study at the University at Buffalo, USA, where she is also Director of the Center for the Study of Psychoanalysis and Culture. The author of *Read My Desire: Lacan against the Historicists* (1994) and *Imagine There's No Woman* (2002), she has also edited a number of book collections, including *October: The First Decade*; *Shades of Noir*; *Supposing the Subject*; *Radical Evil*; *Lacan's Television*; and (with Michael Sorkin) *Giving Ground: The Politics of Propinquity*. Formerly an editor of the art and theory journal *October*, Copjec has for many years been executive editor of *Umbr(a)*, a journal of psychoanalysis.

Elizabeth Grosz is Jean Fox O'Barr Women's Studies Professor of Women's Studies in Trinity College of Arts and Sciences, Duke University, USA. She has most recently published *Chaos, Territory, Art: Deleuze and the Framing of the Earth* (2008) and *Becoming Undone: Darwinian Essays on Life, Politics and Art* (2011).

A. Kiarina Kordela teaches at Macalester College and is Honorary Adjunct Professor at the School of Humanities and Languages, Writing and Society Research Group, University of Western Sydney, Australia. She is the author of *$urplus: Spinoza, Lacan* (2007) – which, beyond linking psychoanalysis to Spinoza and Marx, advances a unified theory of ontology and epistemology – and *Being, Time, Bios: Capitalism and Ontology* (2013) – which addresses more extensively the issue of biopolitics. She is the co-editor of *Freedom and Confinement in Modernity: Kafka's Cages* (2011), and has published essays and articles in collections and journals such as *Spinoza Now* (2011); *European Film Theory* (Routledge, 2008); *The Dreams of Interpretation* (2007); *Modern Language Studies*; *Angelaki*; *Cultural Critique*; *Parallax*; *Rethinking Marxism*; *Political Theory*; *Radical Musicology*; *Umbr(a)*; *Monokl* (in Turkish translation); and *Hihuo kukan [Critical Space]* (in Japanese translation).

Arun Saldanha is Associate Professor of Geography at the University of Minnesota, Minneapolis, USA. He is the author of *Psychedelic White: Goa Trance and the Viscosity of Race* (2007) and co-editor of *Deleuze and Race* (2012) and *Geographies of Race and Food: Fields, Bodies, Markets* (2013). Arun's journal articles and book chapters discuss race, travel, embodiment, Dutch colonialism, drugs, nonrepresentational theory, Marxism and speculative realism. He is working on *Space after Deleuze* and will hopefully soon start writing *The Political Phenotype: Antiracist Science after Man*, a nonhumanist "return to" Darwin and the problem of racial difference in the manner of Lacan's reconceptualization of sex and Althusser's of capital.

Charles Shepherdson is the author of several books on psychoanalysis and contemporary Continental philosophy, and editor of "Insinuations," a book series at SUNY Press. Previously a Member at the Institute for Advanced Study in Princeton, he was a Senior Specialist in American Studies with the Fulbright Program, and National Science Council Visiting Professor in Taiwan from 2006 to 2008. He is currently Professor of English and Director of the Liberal Studies Program at the State University of New York at Albany, USA.

Hoon Song is Associate Professor of Anthropology at the University of Minnesota, Minneapolis, USA, and is the author of *Pigeon Trouble: Bestiary Biopolitics in a Deindustrialized America* (2010). He has recently embarked on a project about how North Korea of late became one of the privileged figures with which liberal democracy's self-understanding is counter-positively organized.

arun saldanha

ONE, TWO, MANY
what is sexual difference now?

tarzan

In the 1918 movie *Tarzan of the Apes* a white baby of noble blood is abandoned in the African jungle after an attack by hostile apes. Kala the mother-ape (apes are played by humans in ape suits) sees in him a substitute for her own prematurely dead infant and raises him with love. Tarzan "did not dream he was different" from the apes until, grown up, he discovers his reflection at the waterfront: fair, hairless, strong, erect. This experience of discovering his "I" is more primordial than what occurs in Lacan's mirror phase, for the difference constituting his imaginary is not only bodily and sexual but evolutionary, racial, ontotheological. The pond scene points at nothing less than the genesis of man in the heart of Africa, and he is already white. The whole film is a comment on secular modernity's deepest anxieties about Darwinism.

Upon recognition of his own species being, Tarzan (who gave him the name is unclear) goes in search of clothing and comes across his parents' skeletons. Out of human instinct he picks up a schoolbook and recognizes the English words "boy" and "monkeys" as well as their meanings. He has entered the symbolic. He will protect civilization from the natives (played by topless African Americans) and wild animals (including jaguars and Indian elephants). He will assert his mastery over the jungle after an exuberant kill of a gorilla. His hut will bear the sign "killer of beasts."

Jane sails to Africa, engaged to a man in an expedition to retrieve the lost aristocratic boy for the health of the metropolis. She is vulnerable not only to wildlife, heat and illness but also her jealous and rapist husband-to-be. After saving her, Tarzan carries Jane away. He "struggles with his shyness" and is clumsy in his efforts to win her love. She says: "Tarzan is a man, and men do not force the love of women." Finally relenting, Jane chooses Tarzan above Europe and renders him fully human through sexual intercourse. After becoming man through imagination and language Tarzan will have to deal with the real of sexual difference, the scary unspeakable longings for a lost species-ambiguous mother which he cannot satisfy in his love for Jane. Their nuptial night in the trees marks the beginning of white settler society in Edenic wilderness.

Lacan has figured large in theorizing sexual difference because his work on the matter is the most intricate and abstract. The colonial myth of Tarzan and Jane nonetheless throws up some questions which could put strain on the Lacanian framework. Why would Tarzan relate to the death of his parents and the rules of culture the same way as a boy in Europe? Could brute violence (he kills ecstatically) be more determining than reflection, speech, and ceremony in his becoming human? Why did cinema turn so early to this highly sexualized origin story just when empires were perilously overstretched? Was this sexualization and racialization not part of a wider shift, as seen in psychoanalysis itself, towards managing sex as core truth of human being?

Is Tarzan an allegory for the humanist narcissism of wildlife preservation? Or does his slinging on vines and desire to both subdue and mimic the lion show he inherited an altogether nonhuman virility? Between ape and human, is there a leap or a continuum? And what about Jane – why did she forsake civilization? From what does the intimacy between her and the cheerful black maid derive? Simply from the fact they're women? There are many dimensions to sexual difference and this introductory essay can only present a selective genealogy of conceptualizing the problem. Lacan will be shown to occupy a central place, even if others are gnawing at his legacy.

man as camera obscura

The Copernican Revolution of Kant defines humanity as the privileged site of truth, while truth is humanity unfolding its essence as perpetually unfinished being. Reason embarks on a quest for a set of rules for all thinking and acting, which are perfectible and inevitable for society to hold together. Inaugurating philosophical anthropology, Kant discerns the revolutionary possibility of thinking the transcendental One of humanity as a progressively self-realizing whole subsuming cultural and dogmatic peculiarities. His famous question in the *Logic,* "Was ist der Mensch?," can only be answered in the abstract. Whether we translate this question as "What is man?" or more recently "What is a human being?" has tremendous implications.[1] The problematic itself seems to violate any strong valuation of the fissures and proclivities constituting the species.[2]

Contemporary theory remains heir to the assumption that the power of human reason suffices to legislate what is good, but largely departs from the claim to universality. For critical theory today, a subject assembles itself only in and across linguistic, geopolitical, libidinal, and technological contingency. All great thinkers of difference, from Marx and Nietzsche to Adorno and Deleuze, made their name to a considerable extent by excavating the sociopolitical paradoxes underpinning Kant's philosophical anthropology.

Hence Luce Irigaray sees in Kant a desperate effort at overcoming the loss of the mother-object and thus of reinstating masculinism at the most supreme level. The sensible realm, imagination, intuition, nature, beauty, extension, the body, the thing-itself: all of these concepts constitute the phantasmal feminine soil for the pure reflection of man. As with Descartes, an entire edifice depends on representation, except that now the subject fully comprehends that *how* the world is reflected within is no divine gift but his own sovereign responsibility to work out. All critical thinking still follows Kant insofar as this imperative to engage the world is acknowledged as its very impetus.

The dissension revolves around how to conceive subjective sovereignty. Irigaray asks:

> Are we to assume that a mirror has always already been inserted, and speculates every perception and conception of the world, *with the exception of itself,* whose reflection would only be a factor of time?... Does the subject derive his power from the appropriation of this non-place of the mirror?[3]

The answer is "evidently." Irigaray laments the arrogance of such a self-founding gesture, typical as it is for men's assumption of and appeal to independent wholeness. Forgetting that woman-soil-nature is the "imaginary sub-basement" for his house of splendid words, man fathoms his mission is to comprehend the correlation between the laws and categories of the universe and their reflection inside. If she had a voice woman would tell him otherwise:

> the action is always inside his house, his mind. And what or who can now put it outside? Only a messenger of revolution perhaps? Or else the fact that this hearth is made of glass and that those glasses – rather tarnished

> by age, their brilliance dimmed, having always in fact been unsilvered or blackened by smoke – mirror so deadly a boredom that, whatever one's firm intent, one might finish by wishing to die – to die of love, were that still possible – rather than have things just go on. Forever.[4]

Man suffocates himself in the prison-house of disembodied cognition, disavowing his love-sickness for the (m)Other and the tediousness of the stories he tells himself. Evidently Kant is no solipsist: the transcendental subject is a formal *function* deriving from the ongoing effort to find its parameters. Per definition it can be embodied in any concrete individual. But crucial is that subjectivity exists only through its distinction from animal nature and mere things.

Quentin Meillassoux argues in vastly different vein that it was Kant's magnificent misstep to consider that relevant reality exists only insofar as it is engaged. Henceforth philosophy would aim at discovering and regulating the correlation between realms separated by the European canon – mind and body, idea and matter, interiority and exteriority – an epistemological inclination Meillassoux calls correlationism. "The great outdoors," the overwhelming magnitude and absolute contingency of the universe itself, is foreclosed for direct understanding.[5]

As Deleuze's more nuanced involvement with the transcendental turn and critique of representationalism shows, Kant in fact introduces exactly the kind of abyssal, "absolutely frightening," fully secular formalism Meillassoux seeks, though it is true he could not follow through the radical conceptual possibilities he had unleashed.[6] What concerns us here is that a range of critiques proclaim a complicity between the Enlightenment's philosophical construct of man-as-humanity and patriarchal Eurocentrism. For Irigaray, Kant has an obsession with erectitude and closure (just a few peepholes filtering what arrives inside) and such is what male subjecthood itself aspires to. Combining Meillassoux and Irigaray, the Great Outdoors can be understood as nothing but femininity at its/her most expansive, feared for its/her immense amorphous plenitude, kept at bay through the very devices of representing it/her. Whereas space is feminized and fixed, time appears as the progressive development of man.[7]

Inspired by Irigaray as well as Merleau-Ponty and Deleuze, Elizabeth Grosz's corporeal feminism retrieves the maelstrom of spacetime, of the ontologically perverse Outside. Ethics and politics emerge not from the interiority of a reasoning and rule-bound mind but *within* the kinetic and affective forces of situated encounters between bodies. A consistent emphasis on bodies guides Grosz from Freud to relativity theory and biology: philosophical anthropology unfurls into a general ontology. We inherit a long tradition from Euclid, Galileo, Newton, and Kant of thinking space and time as absolute coordinates into which life inserts itself. Grosz asks whether another physics can be conceived starting from assumptions men have for centuries been ignorant of:

> in order to conceive bodies, and to understand the kinds of active interrelations possible between (lived) representations of the body and (theoretical) representations of space and time, the bodies of each sex need to be accorded the possibility of a different space-time framework.[8]

What the ongoing debates about Kant's legacy show is that sexual difference is inscribed into philosophy's most basic decisions.

sexual antinomies

A second feminist strategy consists of repeating the transcendental turn to squeeze out its implications for one basic parameter of human existence, namely being sexed. When Lacan says that Kant's morality has as its necessary obverse the contemporaneous fantasies of Marquis de Sade, and that they together invented modern ethics, he asks how reason and law have to be desired through signifiers and are therefore ambiguous and violent to the core.[9] Joan Copjec's essay "Sex and the Euthanasia of Reason" performs a similar short-circuit (a combustively charged because illicit conjoining) between philosophy and psychoanalysis, by anchoring sexual difference in the famous antinomies of the *Critique of Pure Reason*.[10] Copjec affirms Kant's reconstruction of rationality but only as split between its two versions, the masculine and the feminine. These versions do not join into a binary. Sex is two like oil and water are two, two incompatible kinds of being liquid without in-between. Men and women communicate and mingle but the real of their difference is thereby rendered thoroughly opaque.

According to Copjec, then, and unbeknownst to himself, Kant was the first to theorize sexual difference in its *logical* dimension. Kant thinks that at its purest reason is in conflict with itself, structured around four antinomies of two kinds. The first antinomy of time and space and the second of divisibility are the "mathematical" antinomies, concerning fundamental ideas for conceiving the physical world. The third antinomy of freedom and the fourth of the absolutely necessary are the "dynamical" antinomies and they aim at understanding causality. These are four intrinsic deadlocks that reason is fuelled by. Reason is, like sexual difference, productive through its essential failure.

The antinomies emerge from age-old cosmological conundrums. The first consists of the thesis "the world is infinite" at conflict with the antithesis "the world is finite." Kant calls the opposition dialectical, not contradictory. He can hold both in tension insofar as the world-whole is itself a transcendental illusion. Any claims about the world as such, without qualification, are technically false:

> the world does not exist at all (independently of the regressive series of my representations), it exists neither as an itself infinite whole nor as an in itself finite whole. It is only in the empirical regress of the series of appearances, and by itself it is not met with at all.

In order to safeguard the (dialectical) capacity of thought to develop, over time, what initially strikes it as paradoxical (such as the universe's simultaneous finitude and infinitude), Kant makes the radical methodological decision to declare beforehand "absolute totality" impossible.[11]

Copjec's startling gesture is to equate the mathematical antinomies with the "feminine" side of Lacan's sexuation schema (discussed below) and the dynamical with the "masculine." That phenomena can never add up to one object of experience called "world" and that we can nevertheless conceive and theorize the world *as* indeterminate are the two Kantian demonstrations that are formally the same as Lacan's formulas for femininity: no woman is not subjected to the phallic function; not-all of woman is so subjected. Femininity, like the world, does not exist, to the extent that it is known only as nontotalizable series of appearances.[12]

The dynamical antinomies are solved by Kant along a different logical route. First, causality deals not with aggregations of *x*s but with the unbridgeable gap between (noumenal) cause and (phenomenal) effect. Second, dynamics delves into the paradoxes not of the universe but of universality. Kant succeeds in demonstrating the existence of what is universally valid – everything is determined – by paradoxically securing a place for a free transcendental subject within it. There is an iron logic of causation *because* there is a freedom which escapes and precedes it to contemplate it. This counterintuitive line of argumentation is crucial to the recuperation of Kant by psychoanalytical theory. His argument for human freedom simultaneously supplements and subtracts from natural causality. As Slavoj Žižek has it, "logical connection of the phenomena in the universal causal nexus necessarily involves an exception, the noumenal act of freedom which 'sticks out,' suspending the causal nexus and starting a new causal series 'spontaneously,' out of itself."[13]

Of course, such a "first cause" which does not adhere to laws and creates *ex nihilo* is traditionally called God. In Kant it becomes a purely hypothetical element, a nonempirical notion of absolutely free will orbiting the faculty of understanding. To simplify a protracted argument, absolute freedom and the laws of nature can be thought fully compatible by virtue of the fact that the former is *intelligible* though cannot be *known* as is the time-series of experiences.[14] Man's freedom is the exception which proves the rule of nature. To summarize, while incompleteness and indeterminacy underpin the mathematical-female side, exception and universality are central to the dynamical-male side. Where Kant's solution to the dynamical antinomies is tautological – freedom exists as transcendental exception because reason is free to intuit it – this is precisely what for Lacan defines the masculine position: man is nothing but the fundamental lie of transparency and self-legislation.

It may be unclear what justifies equating the Kantian regulation of the thinking of any object x whatever with Lacan's phallic function for which x is, far more narrowly, a speaking subject. Does this not render sexual difference consubstantial with the mind as such, accompanying every possible perception? Yes, sexual difference is indeed transcendental, but only as an inescapable enigma to which one's entire affective and productive life is a singular unsatisfying answer. The question is open whether it is the only such enigma. The importance of Copjec's short-circuiting lies in discovering in Kant a subversive espousal of a nonclassical logic. The cleavages Kant finds at the basis of thought are also what resist treating sexual difference, in religious or functionalist terms, as a relation of complementarity.

A third way in which Kantian philosophy is made to speak to sexual difference is to understand ethics as such as formally feminine. For many feminists the quasi-Newtonian generality of the categorical imperative reflects man's fantasy logic.[15] True ethics breaks with law, utility, and opinion altogether, because it works according to a feminine logic of murky excess which Kant sensed but shied away from making his own.[16] Lacan argues in Seminar VII that all morality, and we can add *a fortiori* moral and analytical philosophy, meticulously aims to prevent contact with what he ominously calls "the Thing" (*das Ding*, a certain specification of Kant's *Ding an sich*), the locus of the primordially lost and unrepresentable first object of love.

Prohibition is in itself masculine not only because it is spoken by the father(s) but because it replays the traumatic separation from maternal flesh. The pathos of courtly love – dragon blocking access to lady, duels to win her love, never-ending nonsatisfaction – exemplifies how in culture the Thing becomes domesticated into stereotypical objects of desire.[17] Morality and beauty are erected to protect man from the horror lying at the core of his unconscious, the incestuous desire for reabsorption into woman which has to be repressed in order to be human at all. The hidden ultimate bliss around which man's drives swerve is so abundantly good that it is evil. Like with the sun, reunion would be fatal. In civilized religion, art, and jurisdiction, therefore, misogyny is intrinsic.

Ethics as opposed to morality does not aim at conformity and reunion, but neither is it anarchic. The prototype of the ethical act Lacan finds in Antigone's seemingly crazy decision in Sophocles' tragedy to defy her uncle and king Creon in the name of sibling love and custom.[18] Custom may ironically be more patriarchal than the law, but the point is that Antigone refuses to give up on her desire to fulfill burial duties for her slain rebel-brother in full consciousness she will be punished by death. Psychoanalytically, Antigone is as woman not terrorized by the incest taboo, which means she can bend the death drive to her advantage. Not the mother-Thing but profane law is unbearable for Antigone.

Lacan points out that her melancholic beauty is an index of that dangerous but fascinating insubordinate part of women. Lacanian theory exalts Antigone's suicidal choice to the most perfect act, perhaps precisely because it cannot be an example. It is a rational choice to sacrifice oneself out of duty, not (primarily) to the state but to a universal sense of justice.[19] Antigone's act has nothing to do with either hysteria or combat. To Lacan, ethics is basically tragic, the essence of tragedy to purify the soul of the "fear and pity" planted in the subject by the symbolic, in order to touch the real, the realm of gods, an experience of limit which cannot but sharply transform the subject.[20]

Alain Badiou will in *Theory of the Subject*, his most Lacanian book, turn Antigone into a topological figuration of the fragile subject of communism: "'Creon' is the name of the superego: the law deregulated – destroyed – by its very own native essence as it returns in excess of the place that it circumscribes. 'Antigone' is the name of anxiety, that is, the principle of the infinity of the real, unplaceable within the regulated finitude of the place," in which "place" is social space as defined by the class-stratified nation-state.[21] The feminine not-all becomes a new avenue for post-Heideggerian ontology, politics, philosophy of language, potentially even formal logic and political economy. The Kantian subject encounters his double, exhibiting not simply some of the more endearing characteristics he had forgotten about, but an onto-logic reconstructing the very One to which he was set to ascend. In this regard Badiou's decision to render technical the symbol ♀ in *Being and Event* is nothing short of monumental.[22] But if ethics as such is feminine, and the law masculine, can there still be a reason and a way to politicize femininity? If feminism merely taps into an excessiveness already characterizing the public sphere as its real, can concrete women create anything altogether new by themselves?

life sexed

Lorenzo Chiesa has emphasized Lacan's "anti-Darwinian" stance: "the subject's disordered Imaginary is, first and foremost, proof of man's – contingently successful – '*dis*-adapted' evolution."[23] Unlike the rest of life which conforms to ecological niches, humans are the only maladapted animal. An original *Hilflosigkeit* (helplessness, awkwardness) necessitates on the infant's part for many years a desire to incite its caretakers' desire. From the moment humans invented speech the baby–mother relationship could no longer be said to be ruled by instinct but by a fundamental misrecognition of one's body in the imaginary and symbolic orders, that is, in rules which operate at odds with the "laws of nature." As Copjec writes, "[t]hat which the individual inherits from her species, her family, her race cannot be located merely in a stateable law or dictate, but includes . . . that excess of the law which cannot be articulated within it."[24]

For Darwin, however, and more so for his vitalist interlocutors, nature itself operates through excess, unpredictability, and disarticulation. Grosz has reread Darwin to imply that what *any* organism can do per definition surpasses what its instincts and circumstances compel it to do.[25] In the West sexuality is now hegemonically understood through a reductionist evolutionary psychology which denies that its takes on Darwin justify male domination and capitalist individualism.[26] We are told that the Darwinian perspective on rape holds that it evolved as a strategy in male hominids to ensure efficient use of the superabundance of their sperm cells. Xenophobia is natural selection's answer to the problem of protecting females and offspring from raids of unfamiliar males.

One response to such survivalist teleology is to make Kant's distinction between the empirical realms of science and the transcendental realms of ethics, art, and psychic topology even more robust. This is what Lacanian theory generally does in the wake of Lévi-Strauss.[27] Another route, taken by John Dupré, Anne Fausto-Sterling, and others, is to defeat biologism on its own turf.[28] We should question the Darwinism of evolutionary psychology and philosophy of mind, which as a rule claim political agnosticism and hence are some of today's most pernicious ideologies. Darwin's own theorizing of sex, race, affect, and contingency can easily be shown to be at odds with what has been made of it.

Central to the critical recuperation of Darwin is his revolutionary understanding of tiny somatic differences as "origin" of species which are, therefore, not the result of divine or taxonomic whim. Grosz has explored the affinities between irreducibly nonutilitarian evolution, complexity theory, and the vitalisms of Nietzsche and Bergson:

> Natural selection entails that the material world, and the other organisms by which a being is surrounded, function as a positive provocation to the self-overcoming that is the most basic characteristic of life, this self-overcoming attesting to the irreducible investment of life in the movement of time, its enmeshment and organization according to the forward direction of time.[29]

Grosz revives the traditionally ignored *The Descent of Man and Selection in Relation to Sex,* which argues that the accumulation of morphological differences does not only occur through natural selection (survival of the fittest) but also through sexual selection (survival of the most attractive).[30] The latter is "costly" for the organism and follows laws entirely different from what Darwin's dominant legacy focuses on. What defines sexual characteristics and copulation is not strategy but beauty, exaggeration, aggression, possibly fatality. Darwin found it more important to elucidate behavioral ambiguities than to reiterate the truisms of adaptation and selfishness as explanatory of life. As for Georges Bataille much later, sexual dimorphism is for Darwin the arena of excessive expenditure.[31] Reducing sexual to natural selection and phenotypic outcomes to genetic predestination, as "ultra-Darwinists" following Richard Dawkins do, is revisionist, ideology rather than science.[32]

Credit goes to all evolutionary theory including Dawkins, however, for doing away with essentialism. Ernst Mayr locates the importance of Darwin in the history of concepts in the scientific emancipation from Platonic and Aristotelian "typological thinking" to "population thinking."[33] For Darwin, individuals were unique and had to be, as species could only have emerged because of variation within breeding populations.[34] Population thinking relates sex directly to diversity since (at least in Mayr's synthetic picture and for sexually reproducing creatures) a species is nothing more than a fuzzy set of sexed bodies exchanging genes to persist in time and space through evolution. Furthermore, Darwin himself would add, mating choice is driven by male performance and female taste, so sex types are not essences either. The ideal peacock is an ongoing evolutionary product, not instigator, of sexual reproduction.

It is increasingly understood, against Dawkins, that evolution does not happen merely to individuals or species but to assemblages of developmental programs, organs, and symbionts.[35] The late Lynn Margulis's research on the DNA of mitochondria (the peripheral energy generators in cells) proves that mitochondria are the vestiges of separate genetic lineages now fully integrated with the lineages of cells and bodies. This means evolution occurs more through cooperation and engulfment than selfishness. There would be no evolution without massive promiscuity, especially amongst bacteria. Sex is the swapping of genetic material before it is something binary: "IN BIOLOGY 1 + 1 DOES NOT EQUAL 2."[36] Seen through complexity theory, sex is an instance of the universe's tendency to combine, mingle, fuse, and expand, that is, of the second law of thermodynamics (everything tends towards disorder).[37]

Even then the emergence of sex remains an enigma because it is not only far more energy intensive than asexual reproduction but also necessitates individuality, motility, and programmed death. Margulis speculates that around a billion years ago a new type of cell replication emerged under seasonal environmental stress, in which one pair of chromosomes was lost. Such so-called meiotic division produces deficient haploid (instead of the usual diploid) cells – the gametes, traditionally named sperm and egg cells. The counterpart of meiosis, the rejoining of chromosomes, is simply fertilization. At some point meiosis became a necessary feature of some societies of cells. These then evolved into distinct bodies and species. In order to facilitate the rejoining of haploid cells, distinct organs developed: the sex organs.

A chapter titled "The Asymmetry of Sex" reminds us that sexual reproduction does not exist without an essential inequality between one gamete and other(s). While DNA is inherited from both parents the organelles of only one parent are replicated in the new individual, with a very small number of exceptions.[38] One gamete, almost always the egg cell, is much larger than the other. Moreover, the sperm's mitochondria are not only fewer but eliminated soon after fertilization in a variety of ways, which shows that sex – the asymmetrical transmission of mitochondria – evolved separately and differently. From an evolutionary standpoint the redundancy of male mitochondria can be called the essential component in sexual differentiation, but exactly why there should be competition at all between the parent cytoplasmic genomes is unknown. Human egg cells have some hundred thousand mitochondria, replicated almost faithfully over hundreds of thousands of years. This allows for tracing ancestry to a "mitochondrial Eve" who lived in East Africa between hundred-and-seventy and two hundred thousand years ago, from whom every human is descended via the maternal line.

Though there are many hermaphroditic species, in most just two distinct versions of adult body came to dominate. Yet the slime mold's oscillation between a moving, feeding mass of asexually reproducing goo and a single-cell stage reproducing sexually through spores is a favorite reminder of how hazy the concepts of body and sex really are.[39] Certain is that, once it appeared, sex afforded more streamlined genetic recombination and greater experimentation with ways of living. Sex triggered an evolutionary dance (what neo-Darwinians call an arms race) between the sexes, each responding to the other's accumulation of features and capacities. Divergence between male and female is checked by physiology and risk, but the almost ridiculous proportions of some antlers and mating rituals amongst some birds show the extent to which utility is trumped by bling.

The question as to why there are only two sexes is as obscure as it is critical. We must first acknowledge the exceptions, especially in the fungi kingdom. The most widespread mushroom, *Schizophyllum commune,* has some twenty-eight thousand mating types, so that reproduction can occur with almost all instead of only 50 percent of other individuals. But for an overwhelming majority of species the number stuck at two, the minimum necessary for recombination. A larger number creates more conflict between mitochondrial genomes whose (mostly nefarious) mutations, moreover, spread more quickly where mating is easy.[40] What is incontrovertible is that the very functioning of and differences between egg and sperm are outcomes of a one-billion-year dance (or tug of war, depending on one's framework). Ecological bottlenecks and sexual selection then brought about the multicellular exuberance of flowers, feathers, and armpit hair.

Unlike in other apes, human females are always "receptive" and their menstrual cycle is not noticeable. This makes all-year pleasure the priority of human sex.[41] But as Tarzan learnt, this is also what makes sex such an insurmountable problem. While our perversions (incest, rape, pedophilia, infanticide, cannibalism, etc.) are common across the animal kingdom, evolutionary psychologists usually rationalize the nuclear family and the sexual division of labor in humans as the most efficient framework for survival *for us*.[42] They willfully ignore Freud's lesson that human groups are fraught with tensions far more intractable than amongst other animals, because humans evolved the densest brains and speech.

Whether morality is an instinct or an achievement of reason, it can never annihilate the erratic movements of desire. Culture itself is constituted by the stress of abiding to, and the possibility of transgressing, taboo. Then again, how relevant is "culture" for understanding the intensities of pheromones, childbirth, orgasm, or itching? When it comes to thinking human difference the major tension to negotiate is that between Kant and Darwin. And Freud lies uncomfortably between.

sex in crisis

Written in spring 1999, Badiou's "Sex in Crisis" summarizes how Freud warned humanity what the loss of the innocence of Enlightenment rationalism would amount to in the twentieth century. Freud exposed the illusions of progress constitutive of patriarchal modernity and the necessity to address them head-on. "With respect to the relations between thought and sex – to what one must indeed call the ineluctable sexuation of the desire to think," in other words, "there is a genuine Freudian inauguration, a founding personal courage exhibited by Freud himself."[43] The discovery of the unconscious entails the realization that war, fear, and uncertainty are ultimately sexual phenomena. The shock of psychoanalysis resides in claiming that neurosis and perversion are questions of degree and circumstance, indeed have to interpenetrate to achieve normalcy: incestuous feelings found sexuality, religion is repression, the law itself breeds crime. Human sexuality contains the potential to contemplate and abrogate it, making Freud wary of the emphasis that biology puts on blind adaptation.[44]

However, while Freud strongly opposed the absorption of psychoanalysis into medicine and the life sciences, the continuities with evolutionary thought should not be belittled.[45]

The analogy drawn between ontogeny and phylogeny by the main popularizer of Darwin in Germany, the Romantically inclined Ernst Haeckel, was a central influence on Freud, whose dodging of the conceptual fallacies and latent racism of this analogy allowed him to suggest that historical myth is recapitulated in individual fantasy.[46] *Totem and Taboo* drew directly on Darwin's idea of the primitive horde to explain how society emerged from the repression of guilt following the murder of a primordial Father.[47] The conceptualization of affect in *Inhibitions, Symptoms and Anxiety* also implicitly superimposed clinical evidence upon Darwin's understanding of the evolution of flight.[48] Though Freud distinguished soma and germ in Weismannian fashion, he notoriously came to defend the Lamarckian possibility of phylogenetic memory, notably of the Jewish people.[49]

Regarding sexual difference, Freud improves on Darwin's preoccupation with dimorphism. Like some theosophical thinkers he proposes that humans are to an extent hermaphroditic and their initial desires unsexed.[50] All humans experience both homo- and heterosexual urges, with parents as first love-objects. Freud concluded his researches thinking that boys and girls are very similarly "phallic" in the earliest stage of their sexual life, when they discover the power of the little penis and the "still smaller clitoris" to create pleasure.[51] He would always hold the assumption, however, that the girl senses the male sex organ is superior to hers (and her mother's).

The father's entry into the nuclear scene involves depriving infants from their primordial sexuality and mother. In the boy there emerges a wish to kill the father in order to reunite with the mother, as the hero does unwittingly in Sophocles' paradigmatic tragedy *Oedipus the King*. He is plunged into a fear of parental law, ultimately based on the fear of being punished by castration. His internalized obedience (the superego) is more severe than the girl's, who understands she *has* nothing, hence nothing to lose. The boy's identification with the father via the Oedipus complex is necessary to usher him into heterosexual masculinity.

The cumbersome route towards "normal" female sexuality – a theme Freud admitted to having great difficulties with – involves first the infantile abandonment of love for the mother in favor of the father (the girl's Oedipalization) and second, as adult, the transfer of erogeneity from clitoris to vagina (orgasm from penetration). Both developments seem to consist in accepting anatomical inferiority in relation to men, whose desire is from the start more straightforwardly heterosexual.[52] Here Freud remained hampered by his nineteenth-century inheritance. Libido theory ("there is only one libido, which serves both the masculine and the feminine sexual functions") and the hypothesis of penis envy ("she has seen it and knows that she is without it and wants to have it") deny any direct experience of female erogeneity.[53] Despite the hermaphroditic hypothesis, Freud thinks femininity transpires only through suppression of original masculinity. Hysteria, frigidity, and melancholia are symptoms of a deficiency in the movements towards father and vagina.[54] Understood as indices of Freud's own limitations as a man in a particular historical geography, his conceptions demonstrate that there are facets of sexual difference he could not grasp. As Lacan will theorize, what structures male authority is the already-deflatedness of the phallus. This includes the authority of Freud and Lacan.

Lacan has shown as no other how phallocentrism entails the profoundly unnatural way the sexes come to be, through a traumatic sense of loss only humans are subjected to: "Freud thus unveiled the imaginary function of the phallus as the pivotal point in the symbolic processes that completes, *in both sexes*, the calling into question of one's sex by the castration complex."[55] In most mammals sexual difference is organized (spatially, libidinally) around the greater physical strength of males, the maternal functions of the females, and ranking behaviors within the sexes. Apes experience helplessness, attachments, violent interruptions by males, etc. But in humans incest between son and mother is rendered quasi-impossible, forming the prototype of law.[56] If male dominance *appears* inexorable it is so for entirely species-specific, partially conscious ways of copulating, talking, birthing, weaning, walking, prohibiting, and praying. The chromosomal is never obliterated by culture and psyche, indeed cannot be expressed outside the latter. Until very recently the universality of patriarchy has been enacted *as if* instinctually, but it is ultimately a chimera. Humans are those apes who evolved to question male dominance.

Melanie Klein further explored our species' intrinsic violence, understanding early childhood as an ordeal fuelled by sadism and cannibalism. The mother's body, especially her nourishing, comforting breast, determines the subject's attitude towards all further objects. Klein agrees with Freud "that the girl wants to have a penis and that she hates her mother for not giving her one." But what the mother "primarily wants is not to possess a penis of her own as an attribute of masculinity, but to incorporate her father's penis as an object of oral gratification."[57] Femininity emerges not so much through the lack of an organ but a *girl-specific* instinctual affection for the parents. The feminine position emerges from a fundamentally different Oedipalization:

> When the female infant gives up her mother's breast and turns to her father's penis as an object of gratification she identifies herself with her mother. But as soon as she suffers frustration in this position too, she very speedily identifies herself with her father, who, she imagines, obtains satisfaction from her mother's breast and entire body, that is, from those primary sources of gratification which she herself has been so painfully forced to relinquish.[58]

Klein still makes the assumption that their sex organs and "introjected penis" make girls inherently disabled. That the boy has a stronger desire for the mother is still presumed to follow from a primordial impetus towards penetration.

However, the disability or passivity of girl/woman can be framed positively. Copjec contends that Freud already hints at the not-all-ness of feminine being.[59] The boy develops a superego which dictates the rest of his life. Forced to posit a mystical plenitude beyond/prior to prohibition, he faces the anxious consequences of the self-contradictions of the law. In contrast, the girl has to accept the groundlessness of the world from the start:

> since the girl's castration takes places without the father's intervention, the (moral) world into which the woman enters appears to her flawed in a different way than it does to the man. For her the fault establishes itself in the absence of a limit, hence a beyond, which absence robs the world of its base or ontological prop.[60]

This is why hysteria is the quintessential feminine illness: it is the compulsion to imagine and resist the law in the relative absence of a superego. This is also why melodrama is the quintessential feminine genre: its realism consists in showing that the world is unmysteriously inauthentic and repetitive.

Copjec's title *Imagine There's No Woman* brings to mind Joan Riviere's influential 1929 paper in which woman is argued to consist *only* of appearance(s):

> Womanliness therefore could be assumed and worn as a mask, both to hide the possession of masculinity and to avert the reprisals expected if she was found to possess it much as a thief will turn out his pockets and ask to be searched to prove that he has not the stolen goods. The reader may now ask how I define womanliness or where I draw the line between genuine womanliness and the "masquerade." My suggestion is not, however, that there is any such difference; whether radical or superficial, they are the same thing.[61]

Masquerade is what makes woman exciting to men, but also frightening, notably in her feminist, hysterical, and *femme fatale* modalities. Ironically, however, women commence their game in masking often in reaction to male anxieties. Nonetheless, the indefiniteness or inexistence of woman is what gives her a "larger," more paradoxical space of possibilities to become-other than what the big Other affords her.

woman and destiny

Psychoanalysis is ambivalent about the emancipation of women. But insofar as it discovers the pervasiveness of sex it is simultaneously indispensable for feminism. Simone de Beauvoir admits sexuality is crucial to the patriarchal organization of society but gives it a phenomenological twist:

> The existent is a sexed body; in its relation with other existents that are also sexed bodies, sexuality is thus always involved; but as the body and sexuality are concrete expressions of existence, it is also from here that their significance can be ascertained: without this perspective [of existence], psychoanalysis takes unexplained facts for granted.[62]

Penis envy and castration are only indirectly relevant, their exaggeration stemming from the perceived privilege of being able to urinate while standing. Instead, Beauvoir sees menstruation as a major event foreshadowing the seemingly inevitable horrors of pregnancy and birthing. The pubescent girl starts discovering with both pleasure and fear that sex is not something private and infantile but a public problem that will sketch her "destiny" (as mother, wife, housewife), a problem which is further intensified because the "sensitivity of the erogenous zones is developing, and they are so numerous in woman that her whole body can be regarded as erogenous."[63]

What psychoanalysis teaches about the split and neurotic subject *The Second Sex* repeats about woman alone. Girlhood is "mutilated," defined by shock, shame, claustrophobia: "She would retain her pride in her bleeding body if she did not lose her self-respect as a human being."[64] Though Beauvoir's descriptions of the inherent aggressiveness of male physiology and coition – "the female is the prey of the species" – led later feminists to accuse her of biological determinism, it follows rather from her existentialist philosophy.[65] For Beauvoir, anatomical difference is versatile matter from which sex is variously assembled.[66] Patriarchy not instinct explains the hugely asymmetrical worlds adolescents find themselves thrown into. The Beauvoir-event lies in showing only half of humanity approximates phenomenology's ideal of the self-projecting, doomed-to-be-free, prehensile subject, whose possessiveness confers a certain thingness onto the other half. Woman is at bottom subjectivity's – man's – objectified Other.

Beauvoir offers a gynocentric response to civilization's censorship. The diffuse, ineffable voluptuousness of female desire is fundamentally at odds with man's approach to orgasm organized as it is in linear, militarist terms. Woman desires the melting of bodies, but she learns to subdue her desire. If women are prone to self-punishment through frigidity, Beauvoir disagrees that women harbor an essential masochism. The generosity underlying female carnality flourishes in lesbianism's recreation of maternal affection. The legitimation of lesbianism in *The Second Sex* as creative choice is pioneering.

Hetero- or homosexual, however, a critical perspective on sex has to be wary of Beauvoir's notion of ecstatic authenticity. Moreover, gynocentrism risks blindness to its own complicity with stereotypes of femininity. Fantasies of Mother Earth and primitive matriarchal paradise are central to New Ageism (think of the spirit mother in the blockbuster movie *Avatar*), representing perhaps the most treacherously patriarchal take on sexual difference. They substitute the analysis of how patriarchy develops within the family and consumer capitalism for the clinically and anthropologically unfounded speculation that a society not governed by the phallus is reachable.

Hominids became humans at the point they lost the innocence of following instincts stamped with their chromosomal make-up, and sex became a confusing mess of taboos, pleasures, and decisions. Feminist utopias and the rare instances of matriarchy are inspiring, but a naturalization of woman as the original giver of life and love suppressed by man's technological violence misses the more radical insights of both ethology and psychoanalysis: the difference between male and female bodies itself does nothing to their propensity to do harm.

women, circulation, signs

Kinship is unthinkable without sexual difference. Claude Lévi-Strauss understood women as objects of circulation between male-dominated groups in order to avoid incest. He binds the theory of the gift of Marcel Mauss with existing ethnography of marriage and the psychoanalytical insight that the incest taboo marks the key difference with other animals. The structuralist approach further hinges on a new formalism borrowed from linguistics.[67] Much material in *The Elementary Structures of Kinship* comes from Australia, where the small size of indigenous "hordes" in comparison to the continent, mostly isolated for thousands of years from other humans, generated particularly clear sexual and kinship practices.[68]

Like Freud, Lévi-Strauss denounces the biological arguments for the incest taboo (that incestuous offspring would be less fit). Instead, exogamy "provides the only means of maintaining the group as a group, of avoiding the indefinite fission and segmentation which the practice of consanguineous marriages would bring about."[69] Not following Lévi-Strauss but Darwin, Grosz has emphasized the essential entwinement of sex and phenotypic diversity. By determining which genes will combine, marriage and cultural ideas of beauty have direct purchase on what the species has come to look like.[70] Under conditions of empire, slavery, and migration as depicted in the Tarzan movies we start speaking of *racial* difference. Race is a globalized system of rules of kinship, that is, of managing the potentialities for variation given with sexual difference, most intensely centered on marriage, but like sex suffusing the spheres of labor, politics, and aesthetics. Unlike sex, however, race is not bifurcation but an ever-changing multidimensional grid.

It is because sexual difference triggers networks of indebtedness that humans have come to diverge culturally and spatially. This spatial imagination brings Lévi-Strauss closer to Mauss and Durkheim and even Marx than to psychoanalysis, which he is quite critical of, especially Freud's own forays into ethnology.[71] Not the law but commerce and generosity define social cohesion, even if the latter are, of course, regulated:

> The prohibition of incest is less a rule prohibiting marriage with the mother, sister or daughter, than a rule obliging the mother, sister or daughter to be given to others. It is the supreme rule of the gift, and it is clearly this aspect, too often unrecognized, which allows its nature to be understood.[72]

Where Freud locates the discharges of sex in a psychic economy, anthropology locates them in a circuitry between families, groups, and nations. Sexual difference is a prime area where anthropology, notably from Margaret Mead onwards, introduced its cultural relativism.[73] Deleuze and Guattari will devote a section of *Anti-Oedipus* to the vexed question of the "universality of Oedipus," concluding that Oedipalization is a hypothetical limit in every social formation, just like capitalism is: the mad beyond to which all circulation of affects and goods *tends*. They thereby push the theorizations of difference of Freud, Lévi-Strauss, and Marx not only together, but raise them to a new philosophical anthropology of flows.[74]

Thus Lévi-Strauss strives, unlike Mead, for an anti-relativist understanding of human relationships as such. Structuralism and psychoanalysis implicitly follow Darwin in the assumption that a human male "ideally" has access to all women in his vicinity, but contra Darwin they insist that this is precisely what requires a leap from instinct. As distant heirs of Kant, structuralism and psychoanalysis both distinguish meaning from nature as its immanent negation. The Lacanian uptake of Lévi-Strauss defends the double notion that only humans signify, but are then trapped in signs. This does not mean signs are everything. If nature could be effectively obliterated by culture, birthing would not necessitate motherhood, (public or suppressed) kinship would be unrelated to genetics, and sex-change operations would present no more psychosocial complications than transvestism does. Corporeal feminism has been at the forefront in examining how the physical aspects of embodiment are always already enfolded into symbolic and imaginary registers.

For psychoanalysis, bodies *cannot* be exhaustively symbolized, imagined, or legislated. Lacan does more than deconstruct the matter/sign contraption. Deconstruction remains on the level of discourse and cannot think *why* the signifier stirs and trails its material enfolding, creating a residue that Lacan calls the real.[75]

In the end the question of nature and culture, so fundamental to the Enlightenment, seems to be as badly posed as it has been productive in generating precision in determining what (of sexuality, of the body) is malleable and what inevitable.[76]

phallus to sexuation

Lacan joins Freud in affirming the ancient idea that there is but one kind of libidinal energy, the masculine. We have seen that Freud argues that the psychic function of the penis is determining for both sexes in the "phallic stage," wherein he thought the girl is largely ignorant of her own genitals. It is true that mythologically and etymologically phallus and penis are interchangeable, but the link is severely abstracted in Lacan. Departing from Freud's biologism-in-the-last-instance he defines phallus as a *pure difference* floating over and above anatomical differences. "In point of fact, phallicism for the girl has always seemed difficult to explain, if not incredible," Lacan concedes. But he continues: "What tends to be overlooked is that it is no less so in the case of the boy. It only *appears* more easy to explain in his case, and I would argue that this apparent ease is a function of phallicism itself."[77] Phallus mediates nature and culture, and it is because it *cannot* coincide with the penis that it is the motor of sexual differentiation at the psychosocial level.

Ultimately, we cannot call the phallus either essentially masculine or neutral. First, it always-already bears witness to the mother and her incarnation in other women. Second, the phallus is not some kind of life force but a signifier, not only compound and ambivalent but effective *as* already lost, as nonpresence, only operational as a veil veiling itself. As Lacan writes:

> Everyone knows that castration is there on the horizon and that it never, of course, occurs. What does happen relates to the fact that the little man is rather a paltry support for that organ, that signifier, and that he seems rather to be deprived of it. And here one can see that his fate is common to that of the little girl, who also can be explained much more clearly from this angle of vision.[78]

The phallic function prescribes that the penis only comes to matter as either absent, or present and removable. But as "master signifier" the phallus hovers around other bodily conditions. What complicates seeing in phallus the final crypto-metaphysical step in man's age-old infatuation with his own virility is that its mastery as signifier is fundamentally tenuous. Containing in Hegelian vein its own annulment, the phallus intrinsically ventures towards void, a pure signifier without signified. Its capacity to stitch all sexual meanings and identities together is based on misrecognition and fantasy. Far from naturalizing phallocentrism, therefore, Lacan calls its bluff. Impotence is the basis, not the accident, of male dominance.

Lacan's essay "The Signification of the Phallus" cannot be understood separately from his announcement that desire is the desire of the Other. In fact, read along Klein, the essay could be read as starting not from the male sex organ at all but from maternal desire. Klein found maternal desire tends towards devouring and reabsorbing the child. This aggressivity opens up in Lacan into a more general conception of the ambiguities of parenting. Freud held that the desire to bear a baby substituted for the lack of a penis, for a sense of social subordination. Lacan adds that motherly love is no given but demanded and maintained by the baby: "The demand for love can only suffer from a desire whose signifier is foreign to it. As the mother's desire *is* for the phallus, the child wants to be the phallus in order to satisfy her desire."[79] Babies generate an "imaginary phallus" (φ) in the relationship with their mother.

Phallus is further explained in Seminar IV on object relations. Boys and girls equally experience the painful endeavor to separate from the mother during weaning. When denied the breast, babies have to accept they are not absolutely desirable and omnipotent. A little later toddlers enter the "dialectic of having/not-having" the phallus. The discovery of the lack in the mother is also the moment the child recognizes that it itself substitutes for the mother's lack. The anxiety of not being the imaginary phallus is stronger in the boy – according to orthodoxy only his desire for the mother is properly incestuous – and he is led to identify with the symbolic phallus (Φ), the network of signifiers first presented to him by the father, in whom he sees his mother's desire heading. The girl can acquire the phallus symbolically as well, but lacking a penis, she is "spared" the fundamental prohibition. What intervenes in any object relation (fetishism and transvestitism being Lacan's cases in point) and identification with a parent is a veil anchored in the phallic function *onto which* fantasies are projected. Hence anatomical difference cannot itself have any direct imprint on the psychic economy even while "the phallus is always behind every relation between man and woman."[80]

"Having" the phallus appears to promise the boy a life of power and love given by mother-substitutes. However, the loss of the satisfaction that the imaginary phallus gave him develops into the castration complex (−φ), an ineradicable sense of having already lost the one satisfaction that matters. Masculinity becomes the violent repression of the mostly unconscious knowledge that the phallus is *not* a biological certitude, into a relentless search for signifiers substituting for this foundational lack. Far from a simple residue of primate life, male dominance in our species is based on an originary, specifically human struggle to forget the trauma of weaning.

Sexual difference names two doomed quests for identity. Girls are in their relation to the maternal phallus the same way as boys, so experience castration when they are subjected to the paternal metaphor. They can continue identifying with the mother in addition to vying for love from the father. They may not "have" the phallus but instead they share the potential to bear children. This insistence on mother–daughter identification is an important improvement on Freud. Still there are the nagging questions as to why the visible detachability of the penis should be foundational and why the girl and mother do not accord their own genitalia any positive signification: "to cut a long story short, for Lacan, the problem with the vagina is simply that is does not 'stick out'."[81] The advent of the concept of sexuation in the early 1970s can be understood as one way out of this residual sexism, paradoxically by following the phallus even further into its logic.

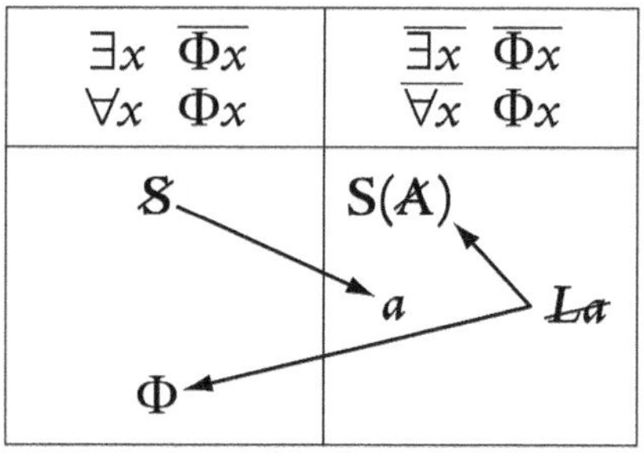

Sexuation is where Lacan's desubstantialized, debiologized materialism becomes most concentrated. His radical Hegelian notion of subjectivity holds that it only exists insofar as the body *thwarts* the subject's emergence and endurance, a resistance to being which Freud called the death drive organized around the impossibility of direct enjoyment.[82] The formalization of the two ways in which humans work out the impasse of selfhood is most fully worked out in Seminar XX (*Encore*), but the three seminars before had been gradually working towards it. Both sexual positionalities thwart Aristotelian logic and Euclidian geometry.[83] On the masculine side there is at least one x who is not submitted to the phallic signifier, whereas it is nonetheless the case for all x that they are thus submitted. The universality of the phallus is strangely secured by an exception, namely the primordial father who is by all men *supposed to* have had access to all women. According to Freud's *Totem and Taboo* "primitive hordes" imagined in a distant past through totemism that they sacrificed and ingested this alpha-male. The unconscious memory of this murder makes possible a fraternity of sons but also renders them guilty and gullible about their own strength. In *Encore* this speculative primatology is stripped to its formal essence: men become men by presupposing one mythical man unencumbered by castration.

On the feminine side there is no x such that it is not submitted to the phallus: there is not a single exception. And yet it is not the case that *all of* x is submitted. "It is thus the very lack of any exception to the Phallus that renders the feminine libidinal economy inconsistent, hysterical, and thereby undermines the reign of the Phallus."[84] On the feminine side there is a universal not-wholeness, while the wholeness of the masculine side is based on the supposition that someone (some one) once escaped its universality. Žižek notes the first row of formulae is existential and particularistic while the bottom row is necessary and universalizing: Lacan oscillates between truths about particular *x*s and truths about *x* as such, as notion.[85] The fourth formula does not read "not all women are phallicized" but "not-all of woman is phallicized." Hence "when any speaking being whatever lines up under the banner of women it is by being constituted as not all that they are placed within the phallic function."[86]

Lacan declines to theorize the connection between nonphallic, "infinite" jouissance and the female sex organs (this will become Irigaray's terrain). What is very clear is that sexual difference is asymmetrical, not complementary: the infamous statement "there is no sexual relation" means each "position" lacks access to what the other side can perceive of its side. In the patriarchal symbolic, Woman is complicit. She is present but as "barred," as not-coinciding with herself, an open set. The *la* of *la femme* is capitalized and crossed to indicate there are only singular women. Man does exist as a universal*ism* – an ideology interpellating concrete bodies – but at the price of greater anxiety, and a generalized perversion.

The Thing, the ultimate object-cause of masculine desire, transforms into *objet petit a,* a surplus of femininity which at once attracts and disturbs masculine subjectivity, while the symbolic phallus Φ is man's surplus attracting and rendering not-whole the woman-position. The diagram underneath the formulae implies that the contradictoriness of the masculine subject derives from its attraction to the residual maternal *a* detectable in women. Woman "inexists" as essence by virtue of a double attraction to the signifier of the big Other's lack (A-barred) and the symbolic phallus. Lacan places the jouissance of the Other-as-lacking – S(A-barred) – on the feminine side because there the big Other does not operate through the law of the exceptional One but through the surplus of the corporeal over itself.[87] Nevertheless, the Other's greatest trick is to lure subjects into thinking there is pure femininity untouched by the phallus. "The assertion 'Woman does not exist' does not in any way refer to an ineffable feminine Essence beyond the domain of discursive experience: *what does not exist is this very unattainable Beyond.*"[88]

In Seminar XXII, *RSI,* a couple of years after *Encore,* Lacan says that man depends on a (technically perverse) fantasy that one special woman exists who he believes can afford him wholeness, and him alone. Lacan's last word on sexuation is that "a woman is the symptom of man." Fairy-tale femininity corresponds to the myth of the primordial father.[89] In its classic versions love is madness to the extent it involves a tragicomic game of deciphering signs to which man is subjected when attempting to possess "his" woman, his Jane.[90] Of course, a woman is *also* what partially escapes her symptomization. A woman's proliferation of surfaces exceeds all the fantasies man (and many women) maintain about her. The feminist upshot is that what is to be enhanced is not supposed substantive qualities like self-sacrifice, intimacy, or irony, since these exist for man, but the surplus virtuality constituted by seeing sexuation afresh from the perspective of the not-all shimmering just outside the reach of signification.

The algebra of sexuation is sufficiently abstract for Žižek to claim that it organizes antagonisms other than the sexual.[91] Hence there is no consensus possible between the subject positions of the bourgeoisie and proletariat. The conflict about how society is structured (either through the universality of inborn freedom or the inconsistency of inequality) is constitutive of society as such, but only Marxism perceives this truth. Communism is structurally feminine: like woman, it is ruled by the phallic function, it accepts that it is historically predated and determined by capitalism, but then posits that its unfinished nature simultaneously offers an illegal, molecular viewpoint that capital cannot foresee. As a transferable topological trope, sexual difference attains world-historical force.

the power of sex

Of all thinkers it is Luce Irigaray who has gone furthest in conceiving female embodiment as potentially autonomous from the phallic function. In Beauvoir the essence of femininity consists of being weighed down by millennia of patriarchy, while her existentialist framework disallows an irreducible split of sex. Following but then breaking sharply with Lacan, Irigaray became the most passionate advocate of sexual difference. If for orthodox Freudianism woman is an appendix to man, albeit one that hurts, Irigaray seeks to restate woman in her forgotten substantive complexity, valorizing her different amorous, economic, and spiritual ways of engaging the world and the other until now systematically obscured. She understands phallus not as originating the sexual field but forced upon it by bodies with penises and epistemologies of penetration, unity, and authority. Girls do not enter society as deficient but enjoy a hidden sexuality, in the final instance predicated on the autoerotic tactility which is/are their vulva. The twoness of her genitalia makes woman the "sex which is not one," the sex which has yet to be.[92] Futurity is generated paradoxically from the eternity of anatomy.

Against Copjec's *Imagine There's No Woman* Irigaray would assert "imagine woman unmediated by man." From a critique of philosophy and the left more encompassing than Beauvoir's, Irigaray in the 1980s moved towards a quasi-Levinasian belief in ethical intersubjectivity in which difference is retained as that which enables. Irigaray is gynocentric: women are better disposed, at the morphological level, to embark on ethical relationships. But all humans, and the life of the planet, stand to benefit from attending to the specificity of women's bodies. What Rosi Braidotti calls sexual difference theory is where a Deleuzian ontology of difference comes to buttress Irigaray's affirmation of fluid, expansive materiality to which women have a special archaic connection.[93] While there is agreement that female sexuality and discourse are to be politicized, a debate was bound to rage on whether positing an autonomous feminine force outside the phallic economy is reactionary or progressive.[94]

Irigaray's lyrical politics of fleshy twoness can be contrasted with the basic parameters of queer theory as laid out in Judith Butler's *Gender Trouble* and *Bodies that Matter*.[95] In order to validate desires and lives outside conventional marriage Butler departs from the metaphysic of manhood-vs.-womanhood. While sexual difference is impasse to Lacan and futurity to Irigaray, to Butler it is the practico-ideological matrix wherein the reproduction of the nuclear family is achieved.[96] The plurality of homo-, bi- and transsexual subject positions demonstrates that the twoness which medicine and the media claim is natural cannot be fought by radicalizing it into the antagonistic impossibility of sexuation, but exploding it altogether into a field of possible performances of gender wherein "sex" loses all of its traditionally presupposed noumenal footing:

> We may seek to return to matter as prior to discourse to ground our claims about sexual difference only to discover that matter is fully sedimented with discourses on sex and sexuality that prefigure and constrain the uses to which that term can be put. Moreover, we may seek recourse to matter in order to ground or to verify a set of injuries or violations only to find that *matter itself is founded through a set of violations*, ones which are unwittingly repeated in the contemporary invocation.[97]

Here, Butler implicitly endorses an extrapolation of body to matter (for there's a lot more matter than that of human bodies) and a reduction of culture to discourse (for bodies are suspended a lot more in material culture than in words). The very category of being is approached only semantically. This echoes Derrida's critique of Heidegger for failing to live up to the aim of overcoming humanism, due to an unspoken denigration of sex in supposedly ungendered terms like being, man, or *Dasein*: "What if sexual difference were already marked in the opening up of the question of the sense of Being and of the ontological difference? And what if, though not self-evident, neutralisation were already a violent operation?"[98]

Freud, Klein, and Beauvoir thought bodily matter becomes imperative through social practices and amassing attendant pleasure, guilt, and fear, but they did not go so far as to say physiological processes like ovulation or ejaculation are *founded* by the violent discourses that regulate them. For Butler, instead of superstructural in relation to the anatomical body which would serve as its ontological bedrock, gender-sexuality is a discursive practice making us *think* such a bedrock exists independently. Sex, moreover, has to be lowered from its psychoanalytical pedestal. Not only are masculinity and femininity on a continuum and are inversions and collages legion, every gender is also inevitably placed in other discursively constructed power relations.

Apart from Derrida, the second key inspiration of gender and queer theory is Michel Foucault's historicization of sexuality, a cornerstone of his larger neo-Nietzschean project complicating the West's trust in truth and progress. Like medicine and criminology with which it interacts, psychiatry is to Foucault a technique for normalizing populations emanating from the urban bourgeoisie.[99] Mainstream psychotherapy attempts to guide its patients towards a standard of self-management, and it by and large accepts middle-class femininity and masculinity as starting points. At the level of populations, biopower raises its head when the reproductive function of sex is acknowledged as essential to the longevity of the nation and the productivity of the state. Slightly earlier than Freud, Havelock Ellis broke new ground regarding infantile sexuality, homosexuality, and the women's movement. But this was exactly what brought him to eugenics. He writes in 1916:

> The breeding of men lies largely in the hands of women. That is why the question of Eugenics is to a great extent one with the woman question [. . .] The breeding of children further involves the training of children, and since the expansion of Social Hygiene renders education a far larger and more delicate task than it has ever been before, the responsibilities laid upon women by the evolution of civilization become correspondingly great.[100]

On the Foucauldian analysis this step from the anatomopolitics of sexology to the biopolitics of whiteness is paradigmatic of the nurturing function of the liberal-capitalist-colonial state.[101] Instead of the dialectical model of Lacan in which the law has to be desired and therefore faces utter contingency at every turn, Foucault understands law as a multiplicity of converging heterogeneous practices which may well incite desire and anxiety in the individual but, more importantly, together delimit for entire territories the limits of what can be done and thought. While broadly sympathetic to Freud's dissection of civilization, Foucault also points out that the discovery of the unconscious dramatically enlarged the scope for the specialized regulation-toleration of sexuality as the supposed nucleus of human existence and happiness.

Although Foucault never acknowledged the import of sexual difference, his argument that power works through bodies and knowledge transformed feminism. But his persistent historicization of "the present" (of the European twentieth century) begs the question as to whether a more general – transcendental, if you will – platform for ethico-political intervention is realizable. Foucault's endorsement of pagan sexual ethics is coextensive with his analysis of its local power.[102] There are no deeper meanings hidden or repressed beneath the networks of practices described. To psychoanalytical theorists, Foucault's "historicism" shares with cultural relativism and social constructionism the problem of ascertaining how ethical thought can justify its movements across contexts. In contrast, Freud has the temerity to theorize the roots of all patriarchy and religion, and Lacan has no qualms drawing the invariant structures of desire.

Gender/queer theory tends to have an exaggerated notion of the power of discourse in determining desire's standards and deviations. As Copjec succinctly criticizes Butler, "the point is that sex is the structural incompleteness of language, not that sex is in itself incomplete."[103] By assuming sex is qua discursive effect unstable, multiple, mobile, etc., one still posits it as something substantial. Sexual difference is *real*, a virtual attractor which by virtue of its necessary (because, ultimately, animal) yet impossible (never experientially or epistemically given) presence prods bodies along. Charles Shepherdson asks:

> Like death, sexual difference is not a human institution, and if in our theories we pretend

> that it is simply one more social construction, invented by a particular society (like democracy or Christianity), do we not unwittingly sustain the humanistic (narcissistic) notion that "man is the maker of all things"? Do we not remain committed to a familiar theory of "man", to the same conception of the "subject" in which sexual difference has always been foreclosed?"[104]

Gender and queer theorists often shy away from theorizing how sexual differentiation is something man has no final control over yet is constantly reinforced by supposedly conscious subjects of all sexes and genders.[105] Without analyzing the ambivalent investments subjects put in their identities and oppression, one will not see that sexual difference is both more than a discursive effect and less than a biological determination. Politically, *Gender Trouble* seems to offer a kind of postliberal voluntarism in the face of discursive determinism. The politics of psychoanalysis – and also of its heir schizoanalysis – derives from appreciating that power is fundamentally vulnerable precisely because it operates through unknowable materiality. Sexual difference and sexuality essentially *escape* society's efforts at subjugating them, yet they demand commitment on the part of therapists, theorists, teachers, and activists to map and minimize their many woes.

The psychoanalytical (or Kantian) claim to universality does not hit a bedrock of reality. Unlike in logical positivism, the formalism of psychoanalysis shows that order *does not* materialize even if it tries to, because the practices we observe are unconsciously designed to erase their deep traumatic origins.[106] Structure is not opposed to the messiness of practice but is already messy, already a paradox.[107] This means that there is, per definition, always room for political subjectification. The virtual structures beneath society are continuously unearthed and transformed through ongoing interpretive work itself not free from desire and ignorance.

Lacan and Foucault need not be antithetical. Shepherdson advocates an ecumenical poststructuralist approach in which Foucault and Derrida bring the topological, linguistic, and historical dimensions of sexual difference into sharper focus.[108] Likewise, queer theory becomes more expedient if its origins in feminism and sexual difference theory are examined. To transform a field as complex as sexual difference requires a conceptual framework built from disparate thinkers, as well as vigilance about the points where patriarchal and racial ideologies slip in through the backdoor.

sexed otherness

Gilles Deleuze is the name for the most systematic attempt to circumvent the Judeo-Christian and German-Idealist traditions out of which psychoanalysis and phenomenology emerged. Deleuze does this by resuscitating the controversies around the philosophies of life in Bergson, Nietzsche, and Spinoza.[109] There has subsequently been a veritable "turn to life" in Continental philosophy, bolstered by the publication of Foucault's lectures on biopolitics, Italian autonomist Marxism's proximity to both Deleuze and Foucault, the resurgence of the animal in post-Heideggerian writing, and the exodisciplinarity of complexity theory. Little of this writing, however, has analyzed sex.[110]

At the heart of Deleuze's problems with Europe's majoritarian philosophical legacy is that it has no immanent concepts of multiplicity and becoming. In Kant and Hegel the progress of knowing unfurls in a human subject which if not an "I" in the Cartesian fashion is at least assumed to be recurrent onto or dividing from within itself. Such an *it-self* logically requires some mode of stable sameness and identity, however paradoxical or dialectical, which instigates and subtends the work of relating to the outside. The elaborations of the dialectic in psychoanalysis, phenomenology, and deconstruction retain the structure of the Two: itself-outside, one-and-another, x-and-its-opposite-already-in-x. In vitalist and evolutionary thought there is often a kind of dualism (time/space, virtual/actual, information/matter, etc.) but only to explain why thinking is activated by the manyness of actual entities interacting and thereby producing novelty.

From the perspective of Deleuze's *Difference and Repetition* the concepts of relationality in Kant and Hegel do not allow for thinking emergence (of a species, an individual, a nation-state) as the immanently generated constraint on the movements of heterogeneous constituent parts.[111] Regardless of how it affects humans, life is unstoppable multiplication organized around thresholds-tensors like sex, predation, and orality, but also earthquakes and the way a fish appears blue by virtue of its green specks. As Grosz has argued in line with biology and implicitly against Lévi-Strauss, nature is not what culture negates but culture is a specification of nature.[112] Deleuzian thought unhinges itself from the prioritization of man and language precisely to complete Kant's critique of the irrationality of anthropocentrism and to live more freely. If *Difference and Repetition* claims a canvas infinitely larger than Lacan's it is not for metaphysical speculation but to define the parameters of synthetic thought in fully Kantian fashion.

The selective recruitment of psychoanalysis and phenomenology for Deleuze's project is usually downplayed.[113] Two extraordinary but disavowed early essays are useful in this regard. In "Description of Woman: For a Philosophy of the Sexed Other," published at twenty and recovered in this journal, we find nothing less than an immanent critique of Sartre for erasing sexual difference, some four years before *The Second Sex*.[114] For Deleuze the phenomenological Other, implicitly a male-Other, is an "expression of a possible world," a phrase borrowed from the manuscript of *Friday* by his friend Michel Tournier. In contradistinction, woman points at nothing transcending herself. "The whole of woman's flesh is consciousness, and all her consciousness is flesh."[115] An influence of Bataille's vivid erotics is noticeable: femininity is luxurious, gracious, even cosmic exuberance (except Deleuze had no conception yet of the death drive as Bataille did). Already under sway of Proust's narratives of sexuality, and prefiguring later conversations with Guattari on the face, Deleuze exalts the bubbly "multiplicity" of woman's freckles above the "irony" of the beauty mark.[116]

Deleuze overstretches himself a little in arguing that the philosopher cannot befriend a woman, because she promises no exteriority, only carnality, "enormous, hot, and full of life." Most men are too timid to deal with such heat. In their very adoration of woman they are prone to a sadist destruction of feminine essence by forcing them to stereotype their secret interiorities. It is instead the indifferent caress that makes love possible even if it does not touch feminine interiority.[117] The essay ends on the fascinating note that the *impure* dynamics of love takes us beyond essences, that is, beyond the existentialist, quasi-Platonic framework wherein Deleuze's thinking commenced. Indeed, his youthful musings on impersonal fragments – ankles, lipstick, frown – already point beyond the philosophical tradition towards the body-without-organs of *Anti-Oedipus*.

The follow-up of a year later, "Statements and Profiles," takes a firmer stand against conceiving otherness as borne by law, friendship, and intellectual transcendence in Socratic or Kantian vein, and by guilt and lack in psychoanalysis. An amoral appreciation of the perversions emerges – notably, via Proust, of the complex sense of love found in pederasty and lesbianism – on a level more philosophical than Bataille. "It is simply a question of becoming a thing, a thing in itself and not a thing for others. And the thing is the unity of contradictories, of the secret and the without-secret, of onanism and exhibitionism."[118] While these two early essays are easily criticized for repeating age-old fantasies of inaccessible feminine essence they also hint at Deleuze's later insight that desire is fundamentally delinquent, presexual, and affirmative, Oedipalized only in the bourgeois framework, which is also the framework of phenomenology and psychoanalysis.

The definitive spillover from sexual difference to general philosophical perversion happens in "Coldness and Cruelty," a long introduction to *Venus in Furs* of Léopold Sacher-Masoch, who "suffered not only from unjust neglect but also from an unfair assumption of complementarity and dialectical unity with Sade."[119] Deleuze gently advocates masochism as a worthwhile becoming for men: emancipation from competitive sadism, enjoyment of an indifferent other, motherly care, patience, and a quasi-alchemical fetishization of certain materials. The severity of the name of the father becomes pleasurable when instilled ritualistically by the dominatrix who derisively flashes the mother-Thing within her.

When psychoanalytical themes appear in *Difference and Repetition* and *The Logic of Sense* sexual difference is never the focus it is for Freud and Lacan.[120] In the former, Deleuze is more interested in theorizing repetition and the death drive, while Klein and Lacan appear in the latter to reconceive becoming through orality and signs. Against psychoanalysis Deleuze posits an elementary and immediate Other-ness of the earth exceeding Lacan's triangulations:

> It is initially in the Other and through the Other that the difference of the sexes is founded. To establish the world without others, to lift the world up (as Friday does [when he blows up the island in Tournier's *Friday*], or rather as Robinson perceives that Friday does) is to avoid the detour. It is to separate desire from its *object*, from its detour through the body, in order to relate it to a pure cause: the Elements.[121]

Sexuality consists of erotic and destructive forces *passing through* bodies, which under colonialist capitalism become anchored in reactionary ideals of masculinity, femininity, and individuality *à la* Robinson Crusoe and Tarzan. Desire does not aim at (and miss) something preexisting but fabricates new immaterial events.

n sexes

Though *Anti-Oedipus* agrees with Lacan that "the phallus is not one sex, but sexuality in its entirety, which is to say the sign of the large aggregate [species] invested by the libido, whence the two sexes necessarily derive," it disagrees that phallus is already castration. Molecular desires operate beyond the sexual binary and even the species:

> the true difference is not the difference between the two sexes, but the difference between the human sex and the "nonhuman" sex . . . Desiring-machines are the nonhuman sex, the molecular machinic elements, their arrangements and their syntheses, without which there would be neither a human sex specifically determined in the large aggregates, nor a human sexuality capable of investing these aggregates.[122]

Interestingly, the idea of a nonhuman sex is obtained from Marx via Lyotard. At some point in his critique of the Hegelian dialectic Marx conjures a difference not encompassed by polarity, complementarity, or opposition.[123] Though he uses sexual difference only as an example, Marx's uncharacteristic quip urges schizoanalysis to demonstrate the continuous slippage of sexual difference by virtue of its ethological, ecological, semiotic mooring:

> Everywhere a microscopic transsexuality, resulting in the woman containing as many men as the man, and the man as many women, all capable of entering – men with women, women with men – into relations of production of desire that overturn the statistical order of the sexes. Making love is not just becoming as one, or even two, but becoming as a hundred thousand.[124]

The point is famously repeated in *A Thousand Plateaus*:

> If we consider the great binary aggregates, such as the sexes or classes, it is evident that they also cross over into molecular assemblages of a different nature, and that there is a double reciprocal dependency between them. For the two sexes imply a multiplicity of molecular combinations bringing into play not only the man in the woman and the woman in the man, but the relation of each to the animal, the plant, etc.: a thousand tiny sexes.[125]

The sexual and the class antagonism are not denied but grasped more precisely as virtual effects of countless everyday practices – not only discursive but financial, nutritional, magical – structured like cobwebs not oppositions. Vitalist ontology is mobilized in a way Bergson or Teilhard de Chardin would find rather unsettling, to scratch away the veneer of naturalness from heterosexist capitalism.[126]

This new materialist view of sex sunders the idealism and closet anthropocentrism from the self–other complex, treating sexual identities as entanglements of bodies, machines, and codes.[127]

The feminist question emanating from Deleuze and Guattari's anti-dialectic revolves around the number of sexes to work with: two or a thousand?[128] It is wrong to think schizoanalysis wants to do away with twoness. When Deleuze and Guattari notoriously posit that all becomings have to pass through a becoming-woman they acknowledge the centrality of sexual bifurcation and conceive femininity akin to the later Lacan and the early Irigaray as a site of incompletion and subversion.[129] But woman is no destiny. The feminist and queer task is to enable a plethora of variations unpoliceable by Oedipal heterosexuality without tumbling into a libertine fantasy world in which women are all too easily reinscribed as playthings.[130]

A more fundamental issue raised by recent critics is whether Deleuze's concept of multiplicity is not false since he has to safeguard it by ontological univocity, an inert *über*-oneness (the Plane of Immanence, anorganic Life, Aîon, etc.) generating and sustaining differentiation.[131] In the sphere of sexuality, the precision of the Lacanian canvas – whose organization around all-too-human themes such as signifier and gaze from the start justifies itself ethico-politically – becomes smudged by the quasi-theological expansiveness of the Deleuzian vindication of life. Deleuze alone oversees how all the tiny sexes explode only to meet again at the zenith of creation. When it comes to psychoanalysis and vitalism, Peter Hallward concludes, "you can't have it both ways."[132] While biology is allergic to any mention of a primeval life force, it is understandable that Deleuze's engagement with vitalism raises the concern in the camp of critical theory too that he is an apolitical pre-Kantian, a closet mystic.

Hence Žižek writes:

> It is crucial to note that *not a single one* of Deleuze's own texts is in any way directly political; Deleuze "in himself" is a highly elitist author, indifferent toward politics. The only serious philosophical question is thus: what inherent impasse caused Deleuze to turn toward Guattari? Is *Anti-Oedipus*, arguably Deleuze's worst book, not the result of escaping the full confrontation of a deadlock *via* a simplified "flat" solution [. . .]?[133]

Žižek does not see Deleuze's politics is so radical that it blinds existing approaches. The imperative of ceaseless becoming, affirming only difference, having done forever with the judgment of God (Artaud), splitting history in two (Nietzsche), is perhaps the most demanding one ever. The Nazi pseudo-revolution and most other occult claims to vitality already betray the principle of becoming in the name of blood, soil, and authenticity. Deleuze's entire oeuvre develops a philosophical solidarity with all those who militate against the strictures of culture, state, personhood, and commodity.

The demand to rid oneself of every fascism was developed most explicitly in *Anti-Oedipus*, the most complete theoretical record of May '68 and an unusual collaboration if ever there was one. What triggered the book was not deadlocks but the suddenly discovered opportunity to augment political and clinical praxis on one side and the refining of concepts on the other – to produce not ontological clamor or flatness but the absolute (impossible) superiority of the molecular, a patient but revolutionary plodding through the ossified left, a return to what was fresh in Marx and Freud, an austere discipline of never ceding to resentment and sadism. To dive into, then cultivate, the nonhuman vectors constituting sexual difference is to actively stave off transcendence, to the precise extent that the differences of life can themselves never allow for such transcendence to occur.

what is sexual difference now?

The problem of sexual difference has lost none of its urgency since the 1918 movie this essay started with. Psychoanalysis's subversive message is all but smothered by the pharmaceutical complex and the mind–body–spirit industry. The media bombard consumers with pornography and imbecilic explanations of sexual behavior. Young people think of feminism as obsolete. Biotechnology promises hitherto unfathomable intervention into the basic parameters of life. Military intervention and anti-immigration measures are justified for battling sexism and homophobia. A silence on the vicissitudes of sexual difference is, like a silence on economic inequality, a succumbing to privileged and dogmatic insolence.

Evidently, understanding even a story as banal as that of Tarzan and Jane leads to a wide array of theoretical options. When stories become banal it only heightens their importance for critical thought. The available tools are heterogeneous and, as befits the problem, not easily reconciled. Two or many, psychoanalysis or vitalism, Kant or Darwin, women's rights or queer performativity, Lacan or Deleuze: is there a way to have it both ways? Unflinching loyalty to one or the other pathway permits extracting a maximum of conceptual force and clarity, at the risk of increasing mutual hostility and incomprehensibility already plaguing the theoretical field. That field is paradoxically also characterized by a joyous eclecticism in which the details of the background from which a notion or term is borrowed are ignored in favor of making an erudite impression.

This essay is certainly a little guilty of the latter. In sketching the outlines of what I think is a crucial juncture in conceptualizing sexuality and gender, however, I distinguished theoretical options starkly and subjectively, so that new conversations across them might transpire. One's approach to theoretical diversity is itself part of an ethos, a particular relationship between desire and discursive regulation. Like in the world at large, by exploring difference and disorder we can at least agree that what we inherit is not good enough.

notes

The origins of this special issue lie in the symposium "What is Sexual Difference Now? A Conversation with Joan Copjec and Elizabeth Grosz" held in Minneapolis on 13 November 2009, with the support of the University of Minnesota's Institute of Advanced Study and Departments of Anthropology, Cultural Studies and Comparative Literature, English, and Gender, Women's and Sexuality Studies, as well as German and Russian Studies at Macalester College. Many thanks to the contributors, journal editors Gerard Greenaway and Charlie Blake, and co-editor Hoon Song, for believing in this project.

1 Manfred Kuehn, "Introduction" to Immanuel Kant, *Anthropology from a Pragmatic Point of View*, ed. and trans. Robert B. Louden (Cambridge: Cambridge UP, 2006) xii.

2 Whether the first difference elided by formal universalism is sexual or economic-racial is a matter of dispute. The colonial, class, and epistemic violence which allowed humanism and the Enlightenment to flourish are well catalogued, as are Kant's own racist contributions to anthropology and geography. The latter cannot be understood separately from the prior definition of reason and morality as the business of European males, which Kant was uninterested in deviating from. On a feminist psychoanalytical interpretation the archaic frustrations of European patriarchy begot the Continent's frantic search for resources and satisfaction elsewhere; sexual difference logically precedes capitalism and racialization insofar as it engages the real, the fundamental prohibition of incest between man and his mother, whereas capital and race operate through networks of signs designed, in the final analysis, to forget prohibition was necessary at all. I tend to follow a Marxist view in which capital and race are themselves reals converging into sexuality.

3 Luce Irigaray, *Speculum of the Other Woman*, trans. Gillian C. Gill (Ithaca, NY: Cornell UP, 1985) 205.

4 Ibid. 213.

5 Quentin Meillassoux, *After Finitude: An Essay on the Necessity of Contingency*, trans. Ray Brassier (London: Continuum, 2008). See my "Back to the Great Outdoors: Speculative Realism as Philosophy of Science," *Cosmos and History* 5.2 (2009): 304–21.

6 Gilles Deleuze, Lecture on Kant at Vincennes University, 14 Mar. 1978, available <webdeleuze.com>. See also Christian Kerslake, *Immanence and the Vertigo of Philosophy: From Kant to Deleuze* (Edinburgh: Edinburgh UP, 2009).

7 See Doreen Massey, "Politics and Space/Time" in *Space, Place, and Gender* (Minneapolis: U of Minnesota P, 1994) 248–72.

8 Elizabeth Grosz, "Space, Time, and Bodies" in *Space, Time, and Perversion: Essays on the Politics of Bodies* (New York: Routledge, 1995) 100.

9 Jacques Lacan, "Kant with Sade" in *Écrits*, trans. Bruce Fink (New York: Norton, 2006) 645–67.

The consequences of this coupling are comprehensively elaborated by Alenka Zupančič in *Ethics of the Real: Kant, Lacan* (London: Verso, 2000), though sexuation is not her main focus.

10 Joan Copjec, "Sex and the Euthanasia of Reason" in *Read My Desire: Lacan against the Historicists* (Cambridge, MA: MIT P, 2002) 201–36; Immanuel Kant, "The Antinomy of Pure Reason" in *Critique of Pure Reason*, trans. and eds. Paul Guyer and Allen W. Wood (Cambridge: Cambridge UP, 1998) 459–550.

11 Ibid. 518–19; bold type removed.

12 Copjec, "Sex and the Euthanasia of Reason" 220–21.

13 Slavoj Žižek, *Tarrying with the Negative: Kant, Hegel, and the Critique of Ideology* (Durham, NC: Duke UP, 1993) 55.

14 Kant, "Resolution of the Cosmological Idea of the Totality of the Derivation of Occurrences in the World from their Causes" in *Critique of Pure Reason* 532–46.

15 See, for example, Monique David-Ménard, "Kant, the Law, and Desire," trans. Leslie Lykes de Galbert, in *Feminist Interpretations of Immanuel Kant*, ed. Robert May Schott (University Park: Pennsylvania State UP, 1997) 341–54.

16 The recuperation in Zupančič's *Ethics of the Real* of an uncanny Kant for which freedom means coming to terms with the subject's internal split from itself, against the polite and wholesome Kant of liberals, is a feminist strategy in this sense. Deleuze similarly came to see in Rimbaud's formula "I is another" a summation of Kant's most difficult lesson: "our interiority constantly divides us from ourselves, splits us into two: a splitting in two which never runs its course, since time has no end." Gilles Deleuze, Preface to *Kant's Critical Philosophy: The Doctrine of the Faculties*, trans. Hugh Tomlinson and Barbara Habberjam (Minneapolis: U of Minnesota P, 1984) ix.

17 Jacques Lacan, *The Ethics of Psychoanalysis*, ed. Jacques-Alain Miller, trans. Dennis Porter (New York: Norton, 1992) 149–51. See also Slavoj Žižek, "Courtly Love, or, Woman as Thing" in *The Metastases of Enjoyment: Six Essays on Woman and Causality* (London: Verso, 1994) 89–112.

18 Lacan himself does not argue that ethics is essentially feminine. One could point to his consistently old-fashioned statements on "man," echoing Heidegger, throughout his seminars, giving rise to grammatical quasi-error:

> Antigone appears as $\alpha\upsilon\tau\theta\nu\theta\mu\theta\sigma$ [autonomous], as a pure and simple relationship of the human being to that of which he miraculously happens to be the bearer, namely, the signifying cut that confers on him the indomitable power of being what he is in the face of everything that may oppose him. (*Ethics of Psychoanalysis* 282)

19 In contrast, when an act goes against every moral duty and even tries to reconceive what duty consists of, it derives from the innate propensity to misuse reason – what Kant calls radical evil. Zupančič, *Ethics of the Real* 82–96. Obviously men are far more prone to evil genius. See also Joan Copjec, ed., *Radical Evil* (London: Verso, 1996).

20 Lacan, *Ethics of Psychoanalysis* 244–48.

21 Alain Badiou, *Theory of the Subject*, trans. Bruno Bosteels (London: Verso, 2009) 162.

22 Idem, *Being and Event*, trans. Oliver Feltham (London: Continuum, 2005). Badiou tempers Lacan's overly enthusiastic adoption of mathematical logic and criticizes his disdain for philosophy. See Russell Grigg, "Lacan and Badiou: Logic of the *pas-tout*," *Filozoksji vestnik* 26.2 (2005): 53–65.

23 Lorenzo Chiesa, *Subjectivity and Otherness: A Philosophical Reading of Lacan* (Cambridge, MA: MIT P, 2007) 7.

24 Joan Copjec, *Imagine There's No Woman: Ethics and Sublimation* (Cambridge, MA: MIT P, 2002) 45.

25 Elizabeth Grosz, *The Nick of Time: Politics, Evolution, and the Untimely* (Durham, NC: Duke UP, 2004); idem, *Time Travels: Feminism, Nature, Power* (Durham, NC: Duke UP, 2005).

26 See, for example, Linda Mealy, *Sex Differences: Developmental and Evolutionary Strategies* (San Diego: Academic, 2000); David M. Buss, *The Evolution of Desire: Strategies of Human Mating*, 4th ed. (New York: Basic, 2003).

27 The most consequential dichotomization ever of nature and history is found in Badiou's *Being and Event*.

28 John Dupré, *Human Nature and the Limits of Science* (Oxford: Oxford UP, 2001); Anne Fausto-Sterling, *Sexing the Body: Gender Politics*

and the Construction of Sexuality (New York: Basic, 2000). Controversies over sex directly lead to controversies over race. See Barbara A. Koening, Sarah S.-J. Lee, and Sarah S. Richardson, *Revisiting Race in a Genomic Age* (Fredericksburg, PA: Rutgers UP, 2008); Anne Fausto-Sterling, "Refashioning Race: DNA and the Politics of Health Care," *Differences* 15.3 (2004): 1–37.

29 Grosz, *Nick of Time* 64.

30 Charles Darwin, *The Descent of Man, and Selection in Relation to Sex* (Princeton: Princeton UP, 1981). An interesting detour into art history is found in George L. Hersey, *The Evolution of Allure: Sexual Selection from the Medici Venus to the Incredible Hulk* (Cambridge, MA: MIT P, 1996).

31 Georges Bataille, *Erotism: Death and Sensuality*, trans. Mary Dalwood (San Francisco: City Lights, 1986).

32 See Richard Dawkins, "Explosions and Spirals" in *The Blind Watchmaker* (Harmondsworth: Penguin, 1986) 195–222; and Matt Ridley's Dawkinsian *The Red Queen: Sex and the Evolution of Human Nature* (London: HarperPerennial, 2003).

33 Ernst Mayr, *One Long Argument: Charles Darwin and the Genesis of Modern Evolutionary Thought* (Cambridge, MA: Harvard UP, 1991) 40ff.

34 It is here, incidentally, that Darwin was so misunderstood by Nietzsche, who had surmised from Haeckel that for biologists species take precedence over individuals.

35 See, for example, Eva Jablonka and Marion J. Lamb, *Evolution in Four Dimensions: Genetic, Epigenetic, Behavioral, and Symbolic Variation in the History of Life* (Cambridge, MA: MIT P, 2006); F. John Odling-Smee, Kevin N. Laland, and Marcus W. Feldman, *Niche Construction: The Neglected Process in Evolution* (Princeton: Princeton UP, 2003); Theunis Piersma and Jan A. van Gils, eds., *The Flexible Phenotype: A Body-Centered Integration of Ecology, Physiology and Behaviour* (Oxford: Oxford UP, 2010).

36 Lynn Margulis and Dorian Sagan, *What is Sex?* (New York: Simon, 1997) 5.

37 Margulis and Sagan, "A Universe in Heat: Sexual Energy" in *What is Sex?* 14–47.

> Producing the order of living copies, by making self-similar living matter, sexual reproduction accelerates the production of disorder because the complex natural organization of a living being inevitably exports more disorder to pay for its organization [. . .] Sex, embedded in animal reproduction, exemplifies the natural tendency for all copying processes to be imperfect. Therefore, sex both protects and disturbs the integrity of living matter, and yet does both in accord with the Second Law of Thermodynamics. (Ibid. 35–36)

On this definition, sex is older than life (ibid. 70ff.).

38 Nick Lane, *Power, Sex, Suicide: Mitochondria and the Meaning of Life* (Oxford: Oxford UP, 2005) 232–41.

39 Margulis and Sagan, *What is Sex?* 105–07.

40 Ibid. 236ff. and the chapter "Why there are Two Sexes" 258–65.

41 See Jared Diamond, *Why is Sex Fun? The Evolution of Human Sexuality* (New York: Basic, 1997).

42

> The human dilemma, then, is that a father and mother must work together for years to rear their helpless children, despite being frequently tempted by other fertile adults nearby. The specter of marital disruption by extramarital sex, with its potentially disastrous consequences for parental cooperation in child-rearing, is pervasive in human societies. Somehow, we evolved concealed ovulation and constant [female] receptivity to make possible our unique combination of marriage, coparenting, and adulterous temptation. (Diamond, *Why is Sex Fun?* 70)

For a critique of biologism from biology see Marlene Zuk, *Sexual Selections: What We Can and Can't Learn about Sex from Animals* (Berkeley: U of California P, 2002).

43 Alain Badiou, "Sex in Crisis" in *The Century*, trans. Alberto Toscano (Cambridge: Polity, 2007) 68.

44 On the relation between Freud and science see Jean Laplanche's *New Foundations for Psychoanalysis*, trans. David Macey (Oxford: Blackwell, 1989) 17ff.

45 See Frank J. Sullaway, *Freud, Biologist of the Mind: Beyond the Psychoanalytic Legend* (New York: Basic, 1979), especially chapter 7; and more recently, Christian Kerslake and Ray Brassier, eds., *Origins and Ends of the Mind: Philosophical Essays on Psychoanalysis* (Leuven: Leuven UP, 2007).

46 The definitive critique of recapitulation, Stephen Jay Gould's *Ontogeny and Phylogeny* (Cambridge, MA: Harvard UP, 1977), shows how Haeckel's biology colludes with vitalism and transcendentalism. Recapitulation had a "pervasive influence" in anthropology, criminology, and psychoanalysis (155–64). A more sympathetic sketch of Haeckelism is found in Robert J. Richards, *The Tragic Sense of Life: Ernst Haeckel and the Struggle over Evolutionary Thought* (Chicago: U of Chicago P, 2008).

47 Sigmund Freud, *Totem and Taboo* [1913] in *The Standard Edition of the Complete Psychological Works of Sigmund Freud*, vol. XIII, ed. and trans. James Strachey (London: Hogarth, 1953–59) (henceforth SE). Darwin, drawing in his turn on ethnography, did not use the term "primitive horde." What Freud noticed was rather Darwin's speculation that hominids already practiced a form of marriage because males were jealous of their wives and offspring; *Descent of Man*, Part II, chapter XX.

48 Sigmund Freud, *Inhibitions, Symptoms and Anxiety* (1925) in SE XX: 126ff.

49 Idem, *Moses and Monotheism* [1923] in SE XXIII.

50

> These long-familiar facts of [anatomical hermaphroditism] lead us to suppose that an originally bisexual physical condition has, in the course of evolution, become modified into a unisexual one, leaving behind only a few traces of the sex that has become atrophied. (Sigmund Freud, *Three Essays on the Theory of Sexuality* [1905] in SE VII: 141)

51 Sigmund Freud, "Femininity" [1932] in SE XXII: 118.

52 Freud's "Female Sexuality" [1931] begins:

> During the phase of the normal Oedipus complex we find the child tenderly attached to the parent of the opposite sex, while its relation to the parent of its own sex is predominantly hostile. In the case of a boy there is no difficulty in explaining this. (In SE XXI: 225)

53 Freud, "Femininity" 131; "Some Psychical Consequences of the Anatomical Distinction between the Sexes" in SE XIX: 252.

54 Later critics of Freud's insistence on penetration for orgasm overlook, however, that in 1905 he thinks "mature" pleasure requires a complex neuronal interaction for which he finds an enticing metaphor:

> When at last the sexual act is permitted and the clitoris itself becomes excited, it still retains a function: the task, namely, of transmitting the excitations to the adjacent female sexual parts, just as – to use a simile – pine shavings can be kindled in order to set a log of harder wood on fire. (*Three Essays on the Theory of Sexuality* 221)

55 Lacan, *Écrits* 463.

56 Amongst both humans and primates the rules around sex are sanctioned through violence. The difference that law makes in our species is not so much that it is learnt and spoken – animals have cultures they communicate – but that it has general instead of group applicability. Cf. Frans de Waal, *Chimpanzee Politics: Power and Sex among Apes*, 25th anniversary ed. (Baltimore: Johns Hopkins UP, 2007).

57 Melanie Klein, *The Psycho-analysis of Children*, trans. Alix Strachey, 3rd ed. (London: Hogarth, 1949) 270.

58 Ibid. 291.

59 Copjec, *Imagine There's No Woman* 115ff.

60 Ibid. 123.

61 Joan Riviere, "Womanliness as a Masquerade" in *The Inner World and Joan Riviere: Collected Papers, 1920–1958*, ed. Athol Hughes (London: Karnac, 1991) 94. The paper is about some intellectual women with lesbian tendencies, but readers have been able to generalize, especially in conjunction with the paper "Jealousy as a Mechanism of Defense," ibid. 104–16.

62 Simone de Beauvoir, *The Second Sex*, trans. Constance Borde and Sheila Malovany-Chevallier (New York: Vintage, 2011) 55.

63 Ibid. 332. "Feminine arousal can reach an intensity unknown by man" (ibid. 411).

64 Ibid. 330.

65 In agitating patriarchy, hyperbole has its place. I would argue for the continuing relevance of Andrea Dworkin, *Intercourse*, 20th anniversary ed. (New York: Basic, 2007).

66 Beauvoir, *The Second Sex* 35. This take on anatomy reverberates in Iris Marion Young, *On Female Body Experience: "Throwing Like a Girl" and other Essays* (Oxford: Oxford UP, 2005).

67 Claude Lévi-Strauss, *The Elementary Structures of Kinship*, rev. ed., ed. Rodney Needham, trans. James H. Bell and John R. von Sturmer (Boston: Beacon, 1969); Marcel Mauss, *The Gift: The Form and Reason for Exchange in Archaic Societies*, trans. W.D. Hall (London: Routledge, 2002).

68 Lévi-Strauss, *Elementary Structures of Kinship* 148ff.

69 Ibid. 479.

70 Grosz, "The Evolution of Sex and Race" in *The Nick of Time* 64–92.

71 Indeed, Lévi-Strauss offers a critique of Freud's Haeckelism in the chapter "The Archaic Illusion" and a lucid denunciation of the assumption behind *Totem and Taboo* that small-scale societies represent an immature stage of development. In the name of universalist synchrony Lévi-Strauss was careful to avoid nineteenth-century evolutionism, effectively naming bourgeois sexuality a desperate exception instead of the presumed rule. *Elementary Structures of Kinship* 84–97, 490–92.

72 Lévi-Strauss, *Elementary Structures of Kinship* 481.

73 Margaret Mead, *Male and Female* [1949] (New York: HarperCollins, 2001).

74 Gilles Deleuze and Félix Guattari, "Psychoanalysis and Ethnology" in *Anti-Oedipus: Capitalism and Schizophrenia*, trans. Robert Hurley, Mark Seem, and Helen R. Lane (Minneapolis: U of Minnesota P, 1983) 166–83.

75 For examples of the deconstructionist feminist critique of nature/culture see Vicki Kirby, *Telling Flesh: The Substance of the Corporeal* (New York: Routledge, 1997); Elizabeth Wilson, *Neural Geographies: Feminism and the Microstructure of Cognition* (New York: Routledge, 1998). For the Lacanian critique of deconstruction see Slavoj Žižek, "The Eclipse of Meaning: On Lacan and Deconstruction" in *Interrogating the Real*, eds. Rex Butler and Scott Stephens (London: Continuum, 2005) 190–212.

76 In *Vital Signs: Nature, Culture, Psychoanalysis* (New York: Routledge, 2000), especially the chapter "The *Role* of Gender and the *Imperative* of Sex," Charles Shepherdson shows how psychoanalysis can stay clear of cultural relativism and its later avatar social constructionism.

77 Jacques Lacan, *Feminine Sexuality*, eds. Juliet Mitchell and Jacqueline Rose, trans. Jacqueline Rose (New York: Norton, 1982) 126.

78 Idem, *Ethics of Psychoanalysis* 308.

79 Idem, "The Signification of the Phallus" in *Écrits* 582.

80 Idem, *Le Séminaire IV. La Relation d'objet*, ed. Jacques-Alain Miller (Paris: Seuil, 1998) 153.

81 Chiesa, *Subjectivity and Otherness* 85. See also Elizabeth Grosz, *Jacques Lacan: A Feminist Introduction* (London: Routledge, 1990) 122ff.

82

> Jouissance, the experience of the body, is produced in a purely contingent manner; it simply happens to the subject. Moreover, if jouissance functions as support of the subject, it is in a way quite different from the being of the philosophers. Jouissance is an unsupportable support, unbearable to the subject who defends against it through the production of a fundamental fantasy. (Copjec, *Imagine There's No Woman* 7)

On desubstantialization see Adrian Johnston, *Žižek's Ontology: A Transcendental Materialist Theory of Subjectivity* (Evanston, IL: Northwestern UP, 2008).

83 See Ellen Ragland, *The Logic of Sexuation: From Aristotle to Lacan* (Albany: State U of New York P, 2004), especially chapter 4.

84 Žižek, *Metastases of Enjoyment* 51.

85 Idem, *For They Know Not What They Do: Enjoyment as a Political Factor*, 2nd ed. (London: Verso, 2008) 121ff.

86 Lacan, *Feminine Sexuality* 144.

87 See Suzanne Barnard, "Tongues of Angels: Feminine Structure and Other Jouissance" in *Reading Seminar XX: Lacan's Major Work on Love, Knowledge, and Feminine Sexuality*, eds. Suzanne Barnard and Bruce Fink (Albany: State U of New York P, 2002) 171–86.

88 Žižek, *Metastases of Enjoyment* 143.

89 Idem, "Woman is One of the Names-of-the-Father, or How Not to Misread Lacan's Formulas of Sexuation," *Lacanian Ink* 10 (1995), available <http://www.lacan.com/zizwoman.htm>.

90 Jacques Lacan, *Seminar XXII: RSI*, unpublished seminar, 1974–75. A translation of the key lecture of 21 January 1975 is found in *Feminine Sexuality* 164–71.

91 See, for example, Slavoj Žižek, "The Real of Sexual Difference" in *Reading Seminar XX*, eds. Suzanne Barnard and Bruce Fink 57–76; "Deleuze and the Lacanian Real," available <http://www.lacan.com/zizrealac.htm>. A. Kiarina Kordela couples sexuation with Marx in *$urplus: Spinoza, Kant* (Albany: State U of New York P, 2007) 99ff. Levi R. Bryant couples sexuation with ontology in *The Democracy of Objects* (Ann Arbor, MI: Open Humanities, 2011) 245ff.

92 Luce Irigaray, "This Sex Which is Not One" in *This Sex Which is Not One*, trans. Catherine Porter (Ithaca, NY: Cornell UP, 1985) 23–33.

93 Rosi Braidotti, "Sexual Difference Theory" in *A Companion to Feminist Philosophy*, eds. Alison M. Jagger and Iris Marion Young (Oxford: Blackwell, 1998) 298–306.

94 See, for example, Maria C. Cimitile and Elaine P. Miller, eds., *Returning to Irigaray: Feminist Philosophy, Politics, and the Question of Unity* (Albany: State U of New York P, 2007).

95 Judith Butler, *Gender Trouble: Feminism and the Subversion of Identity* (New York: Routledge, 1990); idem, *Bodies that Matter: On the Discursive Limits of "Sex"* (New York: Routledge, 1993).

96 Butler later came to an understanding closer to Braidotti's: "sexual difference is the site where a question concerning the relation of the biological and the cultural is posed and reposed, where it must and can be posed, but where it cannot, strictly speaking, be answered," but she remains suspicious of the implied anatomical dichotomy in the term. "The End of Sexual Difference?" in *Undoing Gender* (New York: Routledge, 2004) 186. More on this crucial debate in Rosi Braidotti with Judith Butler, "Feminism by Another Name," *Differences* 6.2/3 (1994): 27–61; Pheng Cheah and Elizabeth Grosz, "The Future of Sexual Difference: An Interview with Judith Butler and Drucilla Cornell," *Diacritics* 28.1 (1998): 19–42.

97 Butler, *Bodies that Matter* 29.

98 Jacques Derrida, "*Geschlecht*: Sexual Difference, Ontological Difference," *Research in Phenomenology* 13.1 (1983) 74.

99 Foucault traces modern sexual biopower to earlier Judeo-Christian and ancient forms in *History of Sexuality Volume I: The Will to Knowledge*, trans. R. Hurley (London: Penguin, 1978); idem, *Psychiatric Power: Lectures at the Collège de France (1973–1974)*, trans. Graham Burchell (Basingstoke: Palgrave, 2006).

100 Howard Ellis, *The Task of Social Hygiene* (Boston: Houghton Mifflin, 1916) 47–48.

101 Michel Foucault, *"Society Must be Defended": Lectures at the Collège de France, 1975–76*, trans. David Macey (London: Picador, 2003) 243.

102 Idem, *The Use of Pleasure: The History of Sexuality Volume 2*, trans. Robert Hurley (London: Penguin, 1985); idem, *The Courage of Truth: Lectures at the Collège de France (1983–1984)*, trans. Graham Burchell (Basingstoke: Palgrave, 2011).

103 Copjec, "Sex and the Euthanasia of Reason" 206.

104 Shepherdson, *Vital Signs* 91.

105 Lacanians argue from the clinical perspective that the phallus cannot be dodged consciously as (some) lesbians try to do, but only indirectly after acknowledging its structuring of all sexuality including their own.

> Female homosexuality fails the not-all through its attempt to eliminate the phallic signifier. [S]he will not have access to feminine *jouissance* [genuinely separate from the phallic function] unless she accepts the "weakness" of being not-all for a man, who can only desire a woman as a partial being, or *objet a*. (Geneviève Morel, "Psychoanalytical Anatomy" in *Sexuation*, ed.

Renata Salecl (Durham, NC: Duke UP, 2000) 37)

Butler conceives of a deliberately contradictory "lesbian phallus" (*Bodies that Matter* 57–92) but forgets that the essential phallic function is to bifurcate two sexes heterosexually. More consistent is then to displace the psychoanalytical edifice entirely; see Grosz, "Reconfiguring Lesbian Desire" in *Space, Time, and Perversion* 173–86.

106 Joan Copjec, *Read my Desire: Lacan against the Historicists* (Cambridge, MA: MIT P, 1994) 12–13.

107 "[I]t is not only that Power itself generates the excess of resistance which it can no longer dominate; it is also not only that – in the case of sexuality – the disciplinary 'repression' of a libidinal investment eroticizes this gesture of repression itself. [Hence] the power edifice itself is split from within: in order to reproduce itself and contain its Other, it has to rely on an inherent excess which grounds it," which Foucault declines to theorize. Slavoj Žižek, *The Plague of Fantasies* (London: Verso, 1997) 26–27.

108 Charles Shepherdson, "History and the Real: Foucault with Lacan" in *Vital Signs* 153–86; idem, "Derrida and Lacan: An Impossible Friendship?," *Differences* 20.1 (2009): 40–86.

109 It is a huge pity that Deleuze did not read more Darwin.

110 Foucault, *"Society Must be Defended"*; Cesare Casarino and Antonio Negri, "It's a Powerful Life: A Conversation on Contemporary Philosophy," *Cultural Critique* 57 (2004): 151–83; Giorgio Agamben, *The Open: Man and Animal*, trans. Kevin Attell (Stanford: Stanford UP, 2003); Jacques Derrida, *The Animal that Therefore I Am*, ed. Marie-Louise Mallet, trans. David Wills (New York: Fordham UP, 2008); Manuel DeLanda, *A Thousand Years of Nonlinear History* (New York: Zone, 1997).

111 Gilles Deleuze, *Difference and Repetition*, trans. Paul Patton (New York: Columbia UP, 1994).

112 Grosz, "The Nature of Culture" in *Time Travels* 42–53.

113 Exceptions include Christian Kerslake, who in *Deleuze and the Unconscious* (London: Continuum, 2007) and *Immanence and the Vertigo of Philosophy: From Kant to Deleuze* (Edinburgh: Edinburgh UP, 2009) demonstrates an indebtedness to both Jung and German Idealism; Monique David-Ménard, who in *Deleuze et la psychanalyse* (Paris: PUF, 2005) understands Deleuze as strengthening the philosophical implications of Lacan's later abstraction; and Joe Hughes, who in *Deleuze and the Genesis of Representation* (London: Continuum, 2008) discovers a Husserlian scheme underpinning Deleuze's entire project.

114 Gilles Deleuze, "Description of Woman: For a Philosophy of the Sexed Other," trans. Keith W. Faulkner, *Angelaki* 7.3 (2002): 17–24. See Jean-Paul Sartre, "Concrete Relations with Others" in *Being and Nothingness: A Phenomenological Essay on Ontology*, trans. Helen E. Barnes (New York: Washington Square, 1956) 471–556, on love, masochism, sadism, etc. For more background see Keith W. Faulkner, "Deleuze *in Utero*: Deleuze–Sartre and the Essence of Woman," *Angelaki* 7.3 (2002): 25–43.

115 Deleuze, "Description of Woman" 18. On the Other-structure see idem, "Michel Tournier and the World without Others" in *The Logic of Sense*, trans. Mark Lester (New York: Columbia UP, 1990) 301–20.

116 Deleuze, "Description of Woman" 20, 21–22.

> A pack of freckles on a face, a pack of boys speaking through the voice of a woman, a clutch of girls in Charlus's voice, a horde of wolves in somebody's throat, a multiplicity of anuses in the anus, mouth, or eye one is intent upon. We each go through so many bodies in each other. (Gilles Deleuze and Félix Guattari, *A Thousand Plateaus: Capitalism and Schizophrenia*, trans. Brian Massumi (Minneapolis: U of Minnesota P, 1987) 35–36)

For Guattari on faciality and Proust see *The Machinic Unconsciousness: Essays in Schizoanalysis*, trans. Taylor Adkins (Los Angeles: Semiotext(e), 2011) *passim*.

117 "In fact, it is the caress that denies all thickness; in tracing a delicate and subtle curve, the caress ceaselessly folds exteriority, draws it into itself, renders it internal to itself" (Deleuze, "Description of Woman" 23). I do not read Deleuze as presenting the caress as futile as Keith Faulkner does ("Deleuze *in Utero*" 34).

118 Gilles Deleuze, "Statements and Profiles," trans. Keith W. Faulkner, *Angelaki* 8.3 (2003) 91.

119 Gilles Deleuze and Léopold Sacher-Masoch, *Masochism: Coldness and Cruelty* and *Venus in Furs*, trans. Jean McNeil (New York: Zone, 1989) 13.

120 Deleuze nonetheless insists on the phallus as the central *problem* disallowing any closure of sexual identity.

> Birth and death, and the difference between the sexes, are the complex themes of problems before they are the simple terms of an opposition. (Before the opposition between the sexes, determined by the possession of lack of the penis, there is the "question" of the phallus which determines the differential position of sexed characters in each series.) It may be that there is necessarily something mad in every question and every problem, as there is in their transcendence in relation to answers, in their insistence through solutions and the manner in which they maintain their openness. (*Difference and Repetition* 107)

121 Deleuze, *Logic of Sense* 317.

122 Deleuze and Guattari, *Anti-Oedipus* 294.

123 Jean-François Lyotard, *Discourse, Figure*, trans. Antony Hudek and Mary Lydon (Minneapolis: U of Minnesota P, 2011) 132–36.

124 Deleuze and Guattari, *Anti-Oedipus* 295–96.

125 Idem, *A Thousand Plateaus* 216.

126 See also what is likely to be one of the best statements on Deleuze's vitalism for a while: Claire Colebrook's *Deleuze and the Meaning of Life* (London: Continuum, 2010).

127 See the pairing of Deleuze–Guattari and Margulis in Luciana Parisi's *Abstract Sex: Philosophy, Bio-technology and the Mutations of Desire* (London: Continuum, 2004).

128 Elizabeth Grosz, "A Thousand Tiny Sexes: Feminism and Rhizomatics," *Topoi* 12 (1993): 167–79.

129

> When Virginia Woolf was questioned about a specifically women's writing, she was appalled at the idea of writing "as a woman." Rather, writing should produce a becoming-woman as atoms of womanhood capable of crossing and impregnating an entire social field, and of contaminating men, of sweeping them up in that becoming. Very soft particles – but also very hard and obstinate, irreducible, indomitable. (Deleuze and Guattari, *A Thousand Plateaus* 276)

See also Rosi Braidotti, "Becoming Woman: Or Sexual Difference Revisited," *Theory, Culture, and Society* 20.3 (2003): 43–64.

130 For all these questions see Rosi Braidotti's *Nomadic Subjects: Embodiment and Sexual Difference in Contemporary Feminist Theory*, rev. ed. (New York: Columbia UP, 2011).

131 Alain Badiou, *Deleuze: The Clamor of Being*, trans. Louise Burchill (Minneapolis: U of Minnesota P, 2000); Peter Hallward, *Out of this World: Gilles Deleuze and the Philosophy of Creation* (London: Verso, 2006).

132 Peter Hallward, "You Can't Have it Both Ways: Deleuze or Lacan" in *Deleuze and Psychoanalysis: Philosophical Essays on Deleuze's Debate with Psychoanalysis*, ed. Leen De Bolle (Leuven: Leuven UP, 2010) 33–50.

133 Slavoj Žižek, *Organs without Bodies: Deleuze and Consequences* (New York: Routledge, 2004) 20–21.

the numbers game

In the mid-1970s a global warming began to melt the icy resistance of feminists to psychoanalysis, thanks to the publication in England of Juliet Mitchell's *Psychoanalysis and Feminism*; the upsurge in France of a group of "New French Feminists"; and the work in the United States of Shoshana Felman, who made a persuasive argument for a feminist-friendly "French Freud." For approximately a decade, psychoanalytic feminism flourished as one of the most exciting and productive discourses of its time. While never completely uncritical of Freudian theory, feminists nevertheless deeply appreciated the fact that it was unique in according a fundamental status to sexual difference and feminine sexuality and thus in making the experiences of women an issue of far-reaching importance, one capable of throwing into question some of the basic assumptions underlying philosophical theories of the subject and political theories of community.

By the mid-1980s, however, signs of a climate change in the relations between feminism and psychoanalysis were already apparent. Teresa de Lauretis telegraphed in the title of her groundbreaking book *Technologies of Gender* the key terms that would oversee the uncoupling of the two discourses, and articulated in her highly prescient preface the slogan under which this uncoupling would effectively take place: "A feminist theory of gender . . . points to a conception of the subject as multiple, rather than divided."[1] The consequences of this formulation and the growing interest it heralded in a rigorous interrogation of psychoanalysis cannot be underestimated. There arose, however, in many of us an uneasy sense that something was being lost in this precipitous embrace of the newly defined category of gender. As "the end goal of the

joan copjec

THE SEXUAL COMPACT

feminist revolution" – defined by Shulamith Firestone as "not just the elimination of male privilege, *but of the sex distinction itself*" – seemed to draw nigh, some wondered whether sexual difference was really as eliminable as class differences were and whether it was desirable to strip the former of significance.[2] In light of this uneasiness I list here the salient and tightly interwoven features of the mid-1980s shift away from sexual difference:

(1) The psychoanalytic category of sexual difference was from this date deemed suspect and largely forsaken in favor of the *neutered* category of gender. Yes, *neutered*, I insist on this; for it was specifically the *sex* of *sexual difference* that dropped out when this term

was replaced by *gender*. Gender theory not only thrust the term *sexual difference* out of the limelight but also it removed the sex even from sex. For, while gender theorists continued to speak of sexual *practices*, they ceased to question what sex is; no longer the subject of serious theoretical inquiry, sex reverted then to being what it was in common parlance: that which is involved in a highly restricted set of activities or in attachments to certain objects or person. Or, within theory, a secondary characteristic that, tired of playing second fiddle, now asserted itself as impudent swagger or naughty voluntarism.

The turn away from the Freudian theory of sex and sexual difference meant that many important questions posed there would also be regarded as outdated and left unattended. Take, for example, the antiquated criticism of Freud's "pansexualism." This charge, that Freud overrated the importance of sex, found it everywhere, the ubiquitous cause of everything, is stunning in its obtuseness. Noting, correctly, that Freud was intent on thinking sex and cause together, his accusers neglected to consider that this reconceptualization of the two in light of each other would leave neither untouched, but would, on the contrary, alter our commonsense notions of both. The Freudian concept of over-determination blurted this fact out but this, too, fell on deaf ears, which heard in the *over* only a surfeit: that is, that the causes of our actions are never unique but always multiple. What ought to have been clear from Freud's exposition is that *over-determination* cannot be adequately approached except as acknowledgment of the subject's *under-determination*. As subjects we cannot trace backward from condition to condition until we arrive at some "lonely hour of the last instance" (as Althusser would later put it) where a cause operated alone to determine our actions. No external or internal necessity guides subjects, who are thus susceptible to endless enticements, to various stimuli, none of which is sufficient by itself to motivate us. It is precisely there where animal instinct is found to be lacking that Freud inserted his speculative concept of drive, which was never a drive to x, y, or z, never connected by necessity to a particular object, for *sex has no domain*. What is essential is not the substitution of a plurality of causes for a single one but the fact that sex as cause cannot be located in any positive phenomenon, word or object, but is manifest in negative phenomena exclusively: lapses, interruptions that index a discontinuity or jamming of the causal chain.

(2) The flight into the multiple, conceived as discrete instances, had, of course, a number of other adverse consequences on the theory of sex. If sexual difference became problematic for gender theory this is because the former was presumed to be heterosexist. It divided subjects into two genres and implied a necessary and/or natural relation between them. (You see what happens when you ignore what I just said: sex emerges as a theory of *under*-determination.) Why – gender theory asks – must there be *only two* genres of persons, two sexes, rather than an infinite number of them? I like to think of this as the Oprah Winfrey theoretical gift of sex: "You get a sex and you get a sex and you get a sex," in which sex can be owned like a car. Refusing this gift, I will instead pose some questions of my own.

Is it automatically the case that many are superior to two? Many are *more numerous*, certainly, but what concerns me is that a precipitous multiplication pushes aside questions that need to be asked, that the proliferation of kinds of subjects (whereby each is her own kind) represents a retreat of thought rather than a theoretical advance. An analogy: Freud conceived the drives as fundamentally antinomic, as divided in two; his enthusiastic contemporaries, however, attempted to "improve" on his theory by multiplying the drives such that every action in which a subject might engage was explained by the existence of a separate drive. (Never mind, again, that drive was never conceived as a drive to x, y, or z.) It quickly became evident, however, that the question of what caused these actions was not answered by the ad hoc proliferation of drives; it was simply deferred. The proliferation of genders repeats this same mistake; it multiplies rather than thinks. That is, it shirks from thinking difference and simply adds another one to a previous one, indefinitely: $1+1+1\ldots$ From where do all these individual ones come? What makes them individual? In large part they come from common-sense observation that there *are* individuals, there are differences, which observation produces an ontological principle (the ontology of the multiple) to be defended, few questions asked. It is simply assumed that an individual comes from herself, that whatever makes a subject this particular subject makes her so per se. This is the nominalist position. Gender theorists operate largely on this assumption.[3]

One sympathizes, God knows, with the reasoning behind their flight into the multiple, their attempt to get out from under an overarching, englobing one in which all differences would be included and greatly reduced to local and minor variations in the nature that unifies them. But this flight does not take us far and it is thus necessary to plot another path. Fortunately, we have at our disposal a philosophical arsenal bequeathed to us by – this will surprise you – the extended elaboration of *the* central concept of monotheism: the concept of the One. The task of the monotheists was to credit – not just theologically, but philosophically – the possibility that one God could serve as the God of all peoples spread across the earth. It is easy to be cynical about this endeavor, to view it as nothing more that a doctrinal mask for the political ambitions of the one Church intent on consolidating its power and gaining dominion over foreign armies and lands. The philosophy of William of Occam and other medieval nominalists provided skeptics of the Church with a razor sufficient to shred this mask to bits. There is, they declared, no other unity than numerical unity, individual beings in themselves and of themselves. No need to posit a separate principle or another reality, for whatever makes an individual man a man makes him *as* concrete individual. There are no universals, no universal man, no species or genera; all such entities lack existence and are simply concepts fabricated by our minds – or, more cynically, by minds intent on gaining power over us by means of these fabrications.

This position was at war with that of Duns Scotus and other realists, who insisted that species and genera were not arbitrary groupings of individuals but *real* entities. Arguing passionately for the reality of universalia, realists insisted on defining their unity as non-numerical. That is, unities, Ones, that could nevertheless not be counted as such inasmuch as they are not determinate but open and non-self-identical. Real universals are for this reason unable to determine the particular nature of any individual and no individual could exhibit the nature of the principle, whether this be the principle of humanity, or God.[4]

Medieval Islamic philosophers contributed to monotheism a compelling conceptualization of the real, non-numerical unity of God, which it expressed succinctly in the formulation "There is no God, but God." God appears in this formulation twice negated. The first negation completely removes Him from the order of living individuals, from human existence.[5] This negation produces the apophatic dimension of God as of all real universals, which are inaccessible to us in their non-determinacy. The second negation announces the appearance of God in the human order, but it does so *without canceling* the first negation. Divine being appears *in* each individual being as that being's innermost core, as the eternal "thisness," the haecceity or Angel of its individuated being; in this way do individuals manifest God, but – again – negatively.[6] This means, as was said, that no individual can exhibit or incarnate God, who is the principle that exceeds the plurality of individuals as well as each, individually. Consider Marx's famous quip that he never once encountered in the streets a universal man, but met there only concrete men. Islamic philosophers and realists in the medieval sense would argue on the contrary that *the universal manifests itself in concrete men* insofar as it forms a part of them. And yet this part is peculiar not only inasmuch as it is greater than the individual of which it is part but also in the sense that it manifests itself negatively as something withdrawn, as unassumable by the individual.

To all individuals subsumed by an abstract universal we can attach predicates that identify them once more: he or she is (in his/her nature) X: *Homo faber*; a political animal; a thinking being. But the real universal does not respond to this model of essences and attributes, or predicates, which undergirds the abstract universal; instead, the former throws such predicates into question. For, if it were true that the nature of God, or man, or Polish people were really present in this person here and could also be present in that person there, we could not truly say – the realists argue – what the nature of God, or man, or Poles is. The real universal withdraws from individuated beings any predicate that might be universally applied to them. Still, the real universal is *not* a fugitive from the One, a flight into the multiple, a skeptic of group belonging. It posits, rather, a fugitive One, a One that flees from itself while multiplying its singular presences. We might say that *the real universal is a living surplus* able to negotiate with historical circumstances, not an abstraction added to on an already existing world.

Now, my proposal is that the real universal exploded in a recharged and expanded form in the Freudian/Lacanian theory of sex. It is through its theory of sex, later also elaborated as drive, that psychoanalysis universalizes human nature as that which has no nature or whose nature is radically plasticized. Devoid of instinct. Or: if in psychoanalysis sex is a universal, it is so in the sense medieval realists gave the term. Sex's paradoxical presence in every subject, as the non-dislodgeable, extimate core of her being, cannot be owned or encompassed by the individual subject; it is a part greater than the individual who "contains" it. And like the real universal, sex is the principle of the subject's individuation. Freud is seldom given the credit due him for preempting the charge that would immediately greet his foray into group psychology – namely, that he was illegitimately trying to extend psychoanalysis beyond the limits of its expertise – when he emphatically insisted at the beginning of *Group Psychology and the Analysis of the Ego* that because a defining tenet of psychoanalysis had always been that the division between the individual and the group *falls within the individual subject*, the group was the proper and legitimate territory of his science. In making this proposal – that the subject was a joint entity, psychic and social at once – Freud effectively articulated the fundamental position of the medieval realists: the principle of individuation is a collective reality present in the individual subject. Indeed, Lacan gave to this reality the name *jouissance* precisely in order to underscore its collective nature. Jouissance, he said, is "an inheritance you can enjoy," you can use, but not "use up."[7] That is to say, the right to this collective reality, to jouissance, is not an exclusive right of any individual; it belongs, rather, to the group.

What matters at this point is this: the One from which the theory of sex starts out is a *real* One, a One that is paradoxical insofar as it is more than itself. This One opposes head on the multiple with which nominalists conduct their love affair. From this starting point, it begins to be evident that the two of sex does not conform to that limited number to which today's nominalists object. Sexual difference is not conceived as a *reduction of the multiple* to a smaller multiple – *only* two – because the two is not just another one, a second one, added to the first. Or: you are mistaken if you think that $1 + 1$ will give you the 2 of sex.

(3) The two of sexual difference was pressured to surrender to the multiplicity of gendered positions in order to respect the historical

variability and constructedness of the subject. Although it was acknowledged that sexual difference was conceived by psychoanalysis not as a biological given but as an effect of a specific technique, or apparatus – namely language – the new wave of feminists worried that the structuralist conception of language was ahistorical and produced effects that were invariant. For this reason the *apparatus* (*l'appareil*) of language was dislodged from its role as the smithy of sex and replaced by historically variable technologies or *dispositifs* – that is, the complex machinery of social practices and knowledges, relations of power, norms and ideals – responsible for constructing gendered positions and relations.

The recourse to *technologies* of *gender* quickly encountered a problem, however: that of technological determinism. How to insure that what came out of the machine was not simply what was put into it, that the gendered subject was not completely stripped of autonomy? This problem was fixed by a well-recognized and anodyne truth: techniques had to be continually redeployed, repeated, but repetition always fails because nothing can be repeated in the same way twice. Or: there is no such thing as repetition.[8] It was on this *denial of repetition* that gender theory staked its hope, for the dooming of repetition meant variation was inevitable and this margin of variation, this slim difference, was seized upon as the site of resistance, the launching pad of thousands of small differences.

The epilogue of *Fear and Trembling* relates an amusing anecdote to which Kierkegaard would implicitly respond in the book immediately following it, *Repetition*. "Heraclitus the obscure" had a disciple who was so inspired by his master's fine thesis that "one could not step in the same river twice" that he was unable to prevent himself from embellishing it further: "One cannot do it even once."[9] Somewhere Lacan, speaking of Heraclitus, refers to the "muddy waters of [the latter's] occultism," thinking perhaps not only of Jung but also of the ancient epithet attached to Heraclitus' name, for Lacan nods in affirmation of Kierkegaard's point in recounting the anecdote of the overzealous disciple. What exactly is Kierkegaard's point? That the disciple inadvertently undermined his master's purpose; for if one remains content with a dismissal of repetition as impossible, one cannot – as Heraclitus intended – affirm movement and change. What the flat denial of repetition obscures is an important fact; if there were no repetition, then the Eleatic *denial* of movement (with which Kierkegaard opens *Repetition*) would be valid. But it is not valid; there is movement, there is change, and these are possible because there *is* repetition.

Gender theory hangs its hat on the impossibility that something could ever be "repeated backward," that is, that an act or experience which had once taken place could take place or be experienced in the same way again. The Greeks called that which gender theorists deny *recollection*. Kierkegaard, however – and Freud, after him – distinguishes recollection from *repetition*, which proceeds in the opposite way by "recollecting forward" an event that had never taken place or a memory that had not "aroused . . . an experience."[10] One of Freud's first examples of this process clarifies what is at issue in repetition as opposed to recollection. Emma suffers from a phobia of entering stores by herself. The origin of the phobia, it turns out, lies not in a single incident but in two incidents taken together. In the first a shopkeeper grabs her genitals through her clothes. An outside observer might say that in this incident she was subjected to sexual assault; but Emma is no outside observer and she herself, too young to know anything about sex, could not and did not experience the assault as sexual. Some time later, having passed the age of puberty, Emma once again enters the store alone. This time two shop assistants laugh at her clothes. While an outside observer would see in this incident no hint of sexual aggression, Emma, who recollects the previous scene forward, experiences that former scene *as if for the first time* and senses a sudden "sexual release."

This canonical example of repetition is also – and significantly – an illustration of what Freud calls the "di-phasic onset of sexuality." What is remarkable about the example is the fact that sex seems to be locatable in neither of the scenes, or snapshots, presented in the analysis. In the first sex is absent from experience, while in the second it is absent from the actions that transpire. One might be tempted to trot out the old pansexualist charge once again or to update it by accusing Freud of perpetrating a "cinematographic illusion" of sorts, not by stringing two still frames together on an abstract, homogeneous timeline to create the illusion of movement but by doing something similar: stringing together two perfectly innocent scenes on the same timeline in order to create the illusion of sex. One need only stop the projection and both – movement and sex – would disappear, like a mirage.

To save both we need to follow the advice of Deleuze and recognize that the instants or "frames" are not static, immobile but rather *mobile sections*, snapshots, precisely, inasmuch as they are *incomplete figures* – "in the process of being formed or dissolving" – of transformation. This simple recognition makes the sequence of snapshots an "immanent analysis of movement," or – in the Emma example – of sex, wherein movement or sex appears as the active link *between* the instants or scenes.[11] This analysis is deemed immanent by Deleuze because it grasps the figures and scenes as they unfold in time, as finite figures and finite scenes. Or, perhaps we should say: because it grasps the finite immanently. We propose this refinement in order to make the point that an immanent analysis regards the finite not as something that is limited to a specific length of time or that is circumscribed chronologically but as what, in its ongoing singularity, has no term and as such repels circumscription.

If the finite, approached immanently, is not defined by a boundary that temporally demarcates it, it nevertheless, and for this very reason, becomes subject to another kind of limit. Not one that cuts it off as a segment from time ongoing, but one that plunges it into its midst. The latter limit injects into finite being a heterogeneity that divides it internally or – better – de-phases it. The finite subject – subject to time – is subject to *delay* rather than to the immediacy of the all-at-once, to a *break*, then, in the all-at-once.

It is important to insist on this point in order to preempt the automatic assumption that intervals or breaks are features only of an *abstract* notion of time, a notion that owes its abstract nature to the fact that it breaks the vital flow of time down into discrete segments of dead time. A non-abstract, immanent notion of time would, it is assumed, restore the continuous flow by eliminating the breaks. In truth, however, the finite subject is not immediately present to a continuous unfolding of events but to breaks, delays, obstacles, still points, to which Freud constantly drew our attention through his invention of a series of concepts, including: a "latency period" that divides the two scenes of sexuality in the Emma case; a "periodic nonexcitability" that interrupts psychic functioning; and a "memory system" that he famously installed between perception and consciousness, thus disjoining them, interrupting their continuity. In his *Project for a Scientific Psychology* Freud describes perceptions as too ephemeral to leave any trace, which means that the perception system remains unsullied, innocent, perpetually ready to receive further impressions, while consciousness is conceived as a belated defense against unconscious memories that have already been recorded. Although this model is altered a bit in "A Note upon the 'Mystic Writing-Pad,'" the disjunction between perception and consciousness retains its prominence and leads Freud to this firmly stated conclusion: "this discontinuous method of functioning of the system *Pcpt-CS.* lies at the bottom of the origin of the concept of time."[12] (Given his early and continued commitment to these models of an out-of-joint time, it is surprising that Freud was ever associated with a theory of continuous biological development.)

The crucial point is this: Freud gives sexuality the same structure that he gives to the temporality of psychic functioning. This relation is not founded on mere analogy; neither term – time or sex – has priority over the other. The two are co-originary. The subject is sexuated inasmuch as she is finite, subject to time. Or: sex belongs not to the essence of the subject but to her historicity; it defines her life of pleasure/unpleasure inasmuch as she is finite, subject to time's vicissitudes.

In the temporal logic of psychic functioning as in the sexual logic brilliantly illuminated by the Emma case, two incidents or moments of time are divided by a break; the second *repeats* the first, but not exactly. This non-coincidence is what triggers the naïve, historicist denial of repetition; "not exactly" is not enough by historicist lights. For Freud, however, things are otherwise; it is non-coincidence, lack of synchrony that repetition repeats. Post-pubescent Emma finds in the earlier scene something – namely sex – lacking, though her discovery is anachronistic, since sex was not lacking *to* pre-pubescent Emma so much as to the distant observer whom the older Emma will come to be. Anachronism – or temporal heterogeneity – is, moreover, doubled, for not only does the past come to be infected by the sense of a displaced present (thus introducing a premature sexuality, arrived too early to be felt) but the present also seems to be infected by a displaced sense of the past (creating a belated experience of sex as a kind of leftover of the former scene). Too early/too late: these are the times of sexuality as well as the times of time itself.

But why not simply see in this a double failure of repetition, rather than a successful repetition, the actual taking place of time and sexuation? Why assert – against the historicist denial of repetition and the Eleatic denial of movement – that the subject does actually become immersed in the waters of sex/time? These questions and denials all arise from the same source: the misguided assumption that breaks and flows are always antithetical. The Emma case belies this assumption. The anachronisms produced by the di-phasic onset of time testify most assuredly to the persistence of a break rather than to a flowing into each other of the two scenes precisely because what is produced is not a homogeneous stream of time. Emma does not make her older self present to what could not have been present to her younger self (there is no sense of a continuous maturation or education here), nor does she reconstitute what *was not* as what *now is*. Each scene is thus internally disrupted as Emma remembers forward what did not yet happen as what had already happened.

Rather than double misfire, however, we witness here the actual onset of sexuality. Emma is sexed. The event of sexuated time passes and to prove it there is a sudden burst – a now – in the sexual release. This now, this burst, happens in a split second, a second that splits rather than gathers the two scenes. Meanwhile, the movement, the passage or flow takes place not between the two scenes but within each. The two scenes in the shop remain the before and after of what divides them and prevents them from flowing together, yet each undergoes an alteration not by the other but in relation to the other. As a result of this, each scene opens up, loses its self-containment. Again we need to caution that this does not mean that one scene comes to contain the other. Instead, both of Emma's encounters – with the store owner in the first case, the shop clerks in the second – become irreducible to the present moment of their taking place. And this is precisely where continuity comes in, finds its footing: for the later scene will find in the earlier one its point of genesis – though this will be not in what happened there but in what did *not* happen.

To respect history is to remain mindful of the fact not only that the past bears on the present but also that the present bears on the past. The two collude with each other, *flow toward each other, but never into each other*. There is temporal continuity but only because there are temporal breaks. The subject is finite, in time, only because she is divided by it, out of synch with it. Staking so much on its denials of division and repetition, gender theory, I would submit, relies not only on an abstract, neutered notion of the subject but also on an abstract notion of time.

foucault enters the mix

Having stated some of my objections to the turn toward gender in the 1980s, I would like to restart the discussion from a different historical moment: the period in the 1920s when heated debates erupted over Freud's theory of castration as essential for the formation of the sexed subject. What many in the fledgling field of psychoanalysis – including Ernest Jones, Helene Deutsch, Melanie Klein, and Karen Horney, among others – found unpalatable was the universality of castration, its indifference to the anatomy of the subjects it was supposed to bring into being. If castration aims at the phallus and the little girl has none, so the reasoning went, then the theory does not do her justice and must be modified to take account of her anatomical and biological differences from the boy. Juliet Mitchell summarizes these early debates in the following way:

> The opposition to Freud saw the concept of the castration complex as derogatory to women . . . Women, so to speak, had to have something of their own. The issue *subtly* shifts from what distinguishes the sexes to what has each sex got of value that belongs to it alone. In this context, and in absence of the determining role of the castration complex, it is inevitable that there is a return to the very biological explanation from which Freud deliberately took his departure.[13]

The first thing to note is that this early opposition to Freud was aimed specifically at his "monotheistic" conception of sex, that is, his thesis that sex and sexual difference could only be thought on the basis of the One.[14] There is only one libido, Freud insisted, and it is male. Abandoning this counter-intuitive thesis like the plague it was, his opponents ended up reducing sexual difference to the pre-linguistic, brute difference between the sexual organs of boys and girls. The second thing to note is that the shift from sex to gender which took place during the debates of the late 1980s resulted in a symmetrical error. The elimination of sexual difference in favor of a study of the social technologies of gender construction left biology behind altogether and produced subjects without any vitality, subjects without bodies or, more precisely, *subjects without sexual organs*.

Given the fact that so much of the work on the social construction of gender relied for its inspiration on Michel Foucault's argument against psychoanalysis in *The History of Sexuality, Volume 1*, a second look at the argument is warranted. Foucault, confronted with student demands for "sexual liberation" during and after the events of May '68, set out in this work to show that this demand for liberation was politically misguided, the rallying cry of a flawed revolt fueled, in significant part, by Freud's "repressive hypothesis." In the face of this harsh accusation, one must be precise about what the father of psychoanalysis *actually* said about repression: he said, specifically, that *ideas* are susceptible to repression and once repressed seek to return into consciousness; but this leaves open the question of whether or not *sex* is repressed. In *The Other Side of Psychoanalysis*, the 1969–70 seminar he delivered in response to these same May '68 demonstrations, Lacan pointed out that Freud's full claim was that in contradistinction to *ideas*, which alone can be repressed, *affect* (or jouissance, in Lacan's vocabulary) is *displaced*.[15] What purpose does this distinction serve? It allows us to observe that affect is not inaccessible to consciousness, does not elude the subject, in the same way as a repressed idea does. For, if there is always a chance that a repressed idea will gain entry and be recognized by consciousness, there is *no chance that jouissance will ever be anything but displaced in relation to consciousness*; it will never find a place that is proper to it. It is this crucial distinction which prompts Lacan's warning that in parading their sexuality the students were in fact "looking for a master," allowing themselves to become the helots of a regime that was pulling their strings.[16] If they sought sanction in Freud's theory of sex for these self-displays, for their attempts to "out" their jouissance, they were knocking at the wrong door. And so was Foucault when *he* attempted to lay a significant portion of blame for the troubling rise of *scientia sexualis* (the hygienic, confessional, let-it-all-out theory of sex) on the doorstep of Berggasse 19.[17] Sex can never be put on display because it *is* nothing other than that teetering, unsettling displacement which permanently throws the subject's identity off balance. In short, Foucault attributed to Freud a position he never held and then attacked it, arguing that far from demanding release from the shackles of power, sex operates in solidarity with it; sex, the notion of sex, Foucault insisted, is saturated with power through and through.

In truth, Lacan and Foucault were *on the same side* in regard to the way sex had – incorrectly – become a political factor during this period and the role it was being made to play in the new paradigm of human domination. Both cautioned the students that the demand for sexual liberation did not oppose power but, on the contrary, played into its hands. What they disagreed on was what sex meant, how it was conceived, in psychoanalysis. Lacan argued forcefully that sex is *not* repressed, that the mechanism of repression does not apply to it, and for this very reason it made no sense to say that sex sought to be liberated from repression. Lacan thus enjoined the students not to sacrifice their enjoyment to those in power by parading it, exposing it *as if it were a predicate – more: the major one – of their identity*. In Foucault's view, sex was nothing more than a fictional construct of power that serves to bind subjects to unified, determinate, and normative identities. Political opposition to bio-power must take the form, therefore, not of liberating suppressed sexual identities but of liberating oneself from them, freeing oneself from classification by their categories. Thus, while Lacan and Foucault were allied in their opposition to the demand for the liberation of sex, on the grounds that this demand was a ruse of power, Lacan put all his energy into showing that sex, or jouissance, was not answerable to the opposition liberation/repression and castigated the jouissance restructured by the demand for liberation as *a sham*, while Foucault pursued the idea that sex and the demand to be liberated, to be known, to assert one's identity, were inextricably intertwined.

But the original historical claim for which *The History of Sexuality* is now best known is this: a mutation took place at the end of the eighteenth century which culminated in what Foucault named at the end of that book "bio-power." The specific mutation that gave rise to this new regime occurred, in his words, in the "mode of relation between history and life." For, while life had previously been viewed as outside history, "in its biological element," it was now *also* placed "inside human historicity, [where it was] penetrated by the latter's techniques and powers."[18] The author of an introduction to Ludwig Binswanger's "Dream and Existence," Foucault endorsed the argument that Binswanger put forward, specifically that "life considered as function [as instinct] is not the same as life considered as history"; the two are by their very nature incommensurable and "it is their incommensurability that justifies the existence of both concepts, each within its own sphere."[19] This "each within its own sphere," the absolute separation of the terms, is placed in jeopardy whenever their incommensurability is ignored, for at this point one of the terms begins inexorably to annex the other. Foucault essentially provides an historical illustration of Binswanger's thesis in *The History of Sexuality* when he argues that bio-power is the annexation of life by power and that this particular denial of the incommensurability of life and history was "an indispensable element in the development of capitalism."[20]

The takeover of vital functions by human history (the latter consisting not only of technologies and power but also language and meaning, everything that constitutes the lived experience of life) is the inevitable result of the "new mode of relation" that effaces the radical distinction between vital functions and lived experience. But because there is, in fact, a radical split, because the terms are incommensurable – as Foucault, following Binswanger, asserts – that which pretends to forge a relation between the terms, or forges a *fraudulent* relation, must itself be fraudulent or, as Foucault puts it, "a mirage."[21] In Foucault's account the "mirage" that allows us to remain blind to the incommensurability of life and human history is precisely sex, inasmuch as the latter which is thus performs a synthetic function:

> [T]he [bio-political] notion of "sex" made it possible to group together, in an artificial unity, anatomical elements, biological functions, conducts, sensation, and pleasures, and it enabled one to make use of this fictitious unity as a causal principle, an omnipresent meaning, a secret to be discovered everywhere.[22]

(Once again the misunderstood notion of pansexualism is treated with contempt.)

Foucault is claiming in effect that *scientia sexualis*, the science of sex, attempted to *make a science of relations*, of the knotting together of incommensurable, disjunct terms. If the fictitious entity, sex, was conceived as a "thing with intrinsic properties and laws of its own,"[23] these properties were those that defined a supposed commonality among otherwise distinct and incompatible terms, and its laws were those that rendered the relations among them predictable. The establishment of a commonality and of predictable relations supported the belief that life could be managed and made to yield greater gains; they also undergird the development of the techniques that put this belief into practice.

This historical thesis, as bold as it is complex, relies nevertheless on an observation that is common enough. It has often been noted that binary oppositions, while purporting to oppose disjunct terms, tend in fact to negate the negating power of one of these terms. Thus neutralized, the second, or unmarked, term loses its independent value and is taken up, sublated, by the first, unmarked term. We saw that in the debates over feminine sexuality that took place in the 1920s, the opposition between biology and symbolic forms collapsed in favor of biology, while the reverse happened in the gender theory debates of the 1980s: biology or life was sublated into symbolic forms and produced de-corporealized subjects or bodies without sexual organs. One scarcely needs to add that the opposition male/female is the best-known example of this, for in this supposed opposition the female term has often been shown to have the value only of a minor exception, one that is easily absorbed by the unmarked, male term that stands in for both. What was new or unique to bio-politics, then, was the invention of something called *sex*, which permitted life to be sublated by history. Prior to this, blood, consanguinity, had played a major role in the machinations of power, but with the invention of the sexual mirage hereditary allegiances and consanguine loyalties tended to be downplayed – if not completely eliminated. The species being promoted by bio-power and the globalized economy of capital came to depend on a more *individualized* notion of the subject, one less encumbered by the older order of hereditary allegiances. The invention of sex, Foucault is saying, aided the construction of a completely individualized notion of the subject, one which caused the realist universal – that in the subject which is more than the subject – to disappear. Or: bio-power issues in the era of the multiple rather than the divided subject.

In an interview he gave on French television just a few years prior to the publication of *The History of Sexuality*, Lacan made a claim so diametrically opposed to Foucault's that it stops one cold: "Back to zero, then, for the issue of sex, since anyway capitalism, that was its starting point: getting rid of sex."[24] (Stated otherwise, as I will argue, Lacan's message is this: capitalism made sex – that in the subject which is more than the subject – disappear.) The television interview aired during the time that Lacan was himself returning to zero, going back to basics in his 1972–73 *Encore* seminar on sex, sexual relation, and feminine sexuality. In retrospect, the entire seminar can be read as a preemptive strike against Foucault's misconstruction of the Freudian problematic of sex. Responsible for triggering much of the French and, later,

Anglo-American feminist interest in psychoanalysis in the 1970s, the seminar is filled with conceptual breakthroughs that were not so much challenged in the 1980s rejection of both psychoanalysis and sexual difference by gender theorists as they were ignored or left un-mined. The formulation for which *Encore* became notorious is the one that stated: "There is no sexual relation [*Il n'y a pas de rapport sexuel*]," even though the meaning of this statement was immediately trivialized, thus rendering it unworthy of the attention it received. The statement was taken to be a refutation of the Hollywood-style fiction of the romantic couple and an effort to expose the fact that actual sexual relations are inevitably freighted with compromise and disappointment, ultimately with failure. The famous observation of Freud, that "men and women are a phase apart psychologically," was embraced as a pessimistic, incontrovertible truth and the admission of failure was celebrated as sober political wisdom.

Still, one has to be a little surprised that the impulse to trivialize did not check itself by pausing to wonder why such dime-store psychology would choose to express itself in this particular way, that is, as a negation of the impersonal phrase *il y a* (there is), a phrase alive with philosophical resonance. In philosophy this phrase is regularly employed to state not something but the fact that this something *is*, that it exists. In its very structure, then, the phrase appears to append to beings a supplement of Being. Everything happens as if the verb *to be* had so atrophied that it required propping up by a prosthetic support of Being. Given this philosophical perspective it becomes clear that Lacan's declaration that there is no sexual relation does not deny that such relations exist, rather it kicks away the prop of Being, which serves to "precomprehend" these relations, turning them into prescriptions or formulas.

If there is no prop or support, no "ontological precomprehension" of being, no Being common to all, this means – in this case – that the sexes are incommensurable, they have nothing, no Being, that is common to them. The notion of their complementarity, which would conceive them as two halves of a common humanity, is thus firmly rejected and Freud's statement about their being out of step with each other psychologically is reconfirmed. One must note, moreover, that the casting aside of prosthetic Being is related to a second gesture which Lacan refers to explicitly as the "lopping off of the predicate."[25] Lacan's text is particularly recondite at this point, although the argument remains intact. Here, with a little more elaboration, is what it says. Refusing in the first gesture the support of a common being, or in refusing to say that "a being is," which would imply that "*its being* is a thing which *is*, that its existence is a thing which exists,"[26] we are thereby permitted to "lop off the predicate," to say, for example, that "man is" without saying what. Now, if "man is" can be considered a complete statement, one that requires no predicate to complete it, this is because the verb "to be" is no longer understood as merely a copula linking a subject to a predicate term. The removal of the prosthesis, which thus allows us to lop off the predicate, testifies on the contrary to the fact that the proper status of the verb "to be" is *verbal*, *active*, to be is to act. To say "man is" without feeling one has to say that he is something or other is to acknowledge that his existence is not a thing, but an act of coming into being. Those who think that Lacan has wandered off the Freudian reservation into some foreign philosophical territory would do well to reread Freud's essay on "Femininity," where the same point is made: "In conformity with its peculiar nature, psycho-analysis does not try to describe *what a woman is* – that would be a task it could scarcely perform – but sets about enquiring *how she comes into being*."[27] How she comes into being, not how she is *constructed* as a woman by society (as this remark has mistakenly be read).

the myth of the third substance

Aristophanes' infamous myth of the two sexes as two halves of a whole forever in search of one another is mocked and dislodged by Lacan in the *Encore* seminar by an antic counter-myth of "the third substance." Lacan begins, seemingly resignedly, by noting: "Nowadays, well, we just don't have that many substances. We have thinking and extended substance."[28] This statement is non-controversially true; Descartes reduced the number of substances to two only and the inheritors of his streamlining have been puzzling ever since over the problem of how to put them together. Having made this anodyne observation, however, Lacan grows more audacious, declaring next that for psychoanalysis two substances are simply not enough. To make up for this deficit, he therefore postulates a third, which he baptizes *enjoying substance* (*la substance jouissante*).

Had he had not preceded this myth with a warning against automatically turning nouns into substances, had he not just effectively argued that being is not a substance but an act, we might have been tempted to think that Lacan was stating here that jouissance is a substance that can be added to the other two to form a link between them, the one that has gone missing at least since Descartes. Lacan would then be performing before our eyes the crime with which Foucault was even then preparing to charge psychoanalysis: the crime of inventing something called *sex* that would function as a mirage, as a vanishing point where the radical incommensurability between what can be thought, experienced, lived historically and the vital functioning of our bodies was obfuscated. But Lacan did try to inoculate us against this misreading of jouissance (his own preferred terms) or of sex, in the Freudian sense. Freud, we know, did claim that the sex of the subject was determinable neither by physical science nor by a psychological study of social behavior; sex cannot be grasped in the Freudian sense either as anatomy or as convention.[29] Neither one nor the other, sex is not, however, a third thing, the missing link that sutures the two.

The counter-myth of the "third substance" challenges the myth of the severed sexes long to be reunited once again, by invoking Freud's own counter-"mythology" of the drive. In a superficial reading, Freud's speculative and widely dismissed "drive theory" would seem less to challenge than to satisfy the longing to reunite what had been torn asunder, and would seem to confirm Foucault's thesis that the concept of sex was just the sort of legerdemain that bio-power needed in order to sublate vital functions into political life. Listen, with Foucault's accusation in mind, to this familiar definition of drive given to us by Freud: drive

> appears to us as a concept on the frontier between the mental and the somatic, as the psychical representative of the stimuli originating from within the organism and reaching the mind, as a measure of the demand made upon the mind for work as a consequence of its connection with the body.[30]

This definition might be misconstrued to suggest that the drive, which occupies the frontier between the psychical and the somatic, is the missing link connecting them to each other.

The myth of the third substance resigns the above misreading of the Freudian drive, however, to the museum of curios and demolishes the notion that sex is a separate, third term that causes the incommensurability of the binary terms to vanish. If we seriously credit Freud's positioning of the drive as archaic, we are obliged in fact to take enjoying substance as *prior* to the "substances" that emerge along its frontier. In other words, the rupture traced by the drive *precedes* and gives rise to that which it ruptures. But sex, or enjoying substance, accounts not only for the radical disjunction of the other two from each other but also for the internal disjunction of each. Sex in this way purloins the substantial, or self-enclosed, dimension of each of the so-called substances. Before proceeding, however, I want to state again as clearly as possible the argument thus far: while Foucault argues that bio-power, abetted by the Freudian theory of sex, eliminates the void between life as function and life as historical experience, or between life and law, and thus eliminates the political space or space of possibility of human action, Lacan argues the opposite: Freud conceives sex as that which takes place in and holds open the space of human action.

If being as such is sexed, if being – defined as an act – is bound up originarily with jouissance, as Lacan maintains, we must look for evidence of this claim in each of the two substances that are "nowadays" assumed to exhaust the field of being. Lacan turns first to thinking substance to examine what becomes of it once Freud appears on the scene peddling his theory of sexuality.

Descartes, who baptized it *res cogitans*, defined the function of "thinking substance" as the formulation of clear and distinct ideas, a fact that makes Freud's instructions to his patients – not that they should think clearly about what troubled them, but that they should, instead, say whatever stupidity popped into their heads – appear scandalous. Indistinct ideas and unsorted nonsense acquire with Freud a value that would have dumbfounded Descartes. But why? In *The Project for a Scientific Psychology* (1896), in a statement that profoundly alters the conception of *res cogitans*, Freud asserted that "it is in relation to a fellow human-being [*Nebenmensch*] that a human-being learns to cognize."[31] Freud's premise is that the occasion of thinking, the incentive for the activity we call thought, is associated with this fellow human being, who was – Freud claims – the first object of our satisfaction. This primal object is, however, enigmatic, for it stays with us forever not as a familiar and fond memory but as a *thing*, a residue that evades judgment.

This thing, this *res*, is the very thing that will desubtantialize thinking substance. As with the classical notion of an underlying substance, this thing, too, is said to stay with us forever, never to abandon us, and thus to be the condition of our permanence or persistence as a thinking subject. And yet, while underlying substance guarantees the subject's self-identity, this thing is, on the contrary, the source of the continuous *aphanisis* of identity, its continuous obliteration. It is a strange fellowship we have with the *Nebenmensch*, for by evading apprehension it refuses to offer any criteria *for* fellowship. This raises a key question, which Freud himself will try to tackle only later in his essay on "The Unconscious." Why do we experience this fellow human being as a fellow, as uncannily close, as inalienably internal, rather than as simply alien? Why do we experience it, Freud asks in that essay, not as a second consciousness, but as we do, that is, as such an intimate (if inassimilable) part of our own consciousness that it can only occur to us as a surplus of ourselves rather than as separate from us? Why do we count ourselves in our difference from ourselves not as two but as one, albeit a paradoxical one, a more-than-one?

We should not let the answer that Freud gives to this question distract us from the radical thinking behind his proposing it in the first place. For not only does Freud make the question proceed from the point that Lacan will emphasize in the *Encore* seminar and elsewhere – that reason, for psychoanalysis, is not divorced from but intimately "concerns jouissance"[32] – he also immediately understand this "first object of satisfaction" as an unsettling of any easy distinction between the other consciousnesses, the community of thinkers, with which thinking puts us in touch and the singularity of our thinking process. By answering that we experience the reality of the unconscious – sex or jouissance – as a part, however alien, of ourselves, Freud posits the "sense of self" as internal to our sense of being in relation to something other and greater than ourselves.

Encore's infamous pronouncement "There is no sexual relation" stands little chance of being understood in isolation from its companion, "*Y a d'l'Un* [there is (some) One]."[33] Sex, we might say, is an *exotic* force; it pushes us away, or severs us from ourselves. Sex is in this sense the enemy of relation conceived as unbroken link (just as it separates the sexes from each other) as much as it is the enemy of a certain One, a unifying One. It is for this reason that Freud found himself unable to countenance the existence of what his friend Romain Rolland described as an "oceanic feeling," that is, "a feeling of an indissoluble bond, of being one with the external world."[34] Rather than a feeling, Rolland's description of the oceanic struck Freud as being "of the nature of an *intellectual perception*."[35] Like all abstractions, Rolland's notion of an oceanic oneness carries no conviction of existence; nothing in the concept necessitates its existence. Refusing, therefore, to acknowledge its existential validity, Freud turned to his own theory to show how such an illusion might mistakenly claim to find support there, first of all in its discrediting of the purported autonomy of the ego. Specifically, the psychoanalytic claim is that the boundaries between the ego and its outside are weakened insofar as the ego is "continued inward, without any sharp delimitation, into an unconscious entity . . . designated as the id."[36] Freud implies, without pressing the point – no doubt out of a desire to find some common ground between his ideas and those of his friend – that this weakening of the ego's boundaries does not authorize the existence of an oceanic feeling. Why not? The answer lies in the problematic "continuity" Freud too quickly asserts. For: between the unconscious and consciousness, between the ego and the id, *there is no continuity* (in the common sense); there is only discontinuity, disruption, displacement. The "inner sense" I have of myself as other than myself – thanks to the unconscious and to the "first (ungraspable) object of satisfaction" – precisely *prevents* the oceanic feeling, the feeling of being *immersed* in the One, from overtaking us. Countless paradoxes attest to this primary psychoanalytic insight, inscribed in Lacan's formulas of sexuation as the antinomic relation between the universal and existential quantifiers, the all and its exceptions. This antinomy is also acknowledged by Freud when he insists that each of us unconsciously denies his own mortality even though we accept the elemental fact that man is mortal. Individual subjects view themselves as exceptions, albeit exceptions that do not contest the validity of "the One" to which we belong. "I" and "we" are forever antinomic, in tension by definition, though this tension does not amount to a contradiction: to an "I" *or* "we." This is further confirmation of the point we have been making all along: the One of psychoanalysis is paradoxical, pre-numerical, a One that is more than – and simultaneously, as we will see – less than one.

compactness: the erogenous zone

Zeno's paradox of Achilles and the tortoise enters into Lacan's discussion of sexual difference in *Encore* as if to illustrate the radical differences that divide men and women: swift vs. slow; incremental half-steps vs. continuous movement; man vs. beast. But this somehow misses the point, for however different Achilles and the tortoise are from each other, they do not simply go off in their own directions, diverge. On the contrary, they constitute converging series, since each progresses toward the same limit, which they eventually reach – even if they do not meet there. At the limit Achilles will not catch up with the tortoise, but surpass her. So much for our romantic hopes of an encounter, an ecstatic fusion! Instead, the two remain a phase apart, retain their differences from each other. What, then, is the point of noting their convergence?

Since Zeno we have been accustomed to conceiving the limit only negatively, as unreachable, as defining an impossibility of movement. Yet Deleuze, in his book on Leibniz, speaks of convergent series, which tend toward a limit but do not always possess a final term, in positive terms as entailing *intensities*.[37] Convergent series, which tend toward a common limit without meeting because there always remain an infinite number of points between them, create a positive condition which has a technical name: *compactness*. In *Encore*, Lacan invokes this condition: "I will posit here the term 'compactness.' Nothing is more compact than a fault."[38] What Lacan refers to as a "fault" is what we have been calling a limit. A fault or limit defines a locus or tight space, as Lacan acknowledges when he asserts that "the space of jouissance . . . proves to be compact."[39] A compact space, we could therefore say, is an erogenous zone, or: compactness serves here to explicate the notion of erotogeneity. Lacan makes this argument more or less explicitly when he speaks in a rather strange way about a bed, as if he were not talking merely about a mundane object but a psychoanalytic concept. Indeed, we must read this seminar as his conversion of a bed, as space of erotic encounter, into a concept by describing it precisely as compact, a space in which two people "squeeze each other tight," or: experience jouissance.[40]

The first thing to note is that compactness is a space of impossibility, the impossibility of union or encounter, *and at the same time* a space where something out of the ordinary happens: jouissance. This draws attention to a truth on which psychoanalysis has always insisted: sexual enjoyment – jouissance – emerges from an encounter with the impossible, it *depends on* a limit. Freud, for example: "an obstacle is *required* in order to heighten libido."[41] What prevents this Freudian insight's reduction to the *psychological* observation that libido is ignited by an obstacle – whether mounted by the other's coyness, social or familial taboo, or mere happenstance – that impedes the subject's amorous approach to the other is the fact that in the *psychoanalytic* understanding the obstacle (the limit or impossibility) separating the subject from the other is not breached or dissolved but remains fully intact. It (that is, the impossibility itself) and not its overcoming ignites jouissance.

This point can be brought back to our earlier reference to the *Nebenmensch*, the Thing, mentioned in *The Project*. Freud inadvertently blocks our theoretical curiosity and thus our understanding of this concept by defining it prematurely in the text as the first object of our satisfaction. According to the theory he will develop here and elsewhere, however, there is no object of satisfaction *before* the loss of the object, no mother before her withdrawal. Thus, it is not that the first object of our satisfaction is later lost, but that this object is lost before it exists as an object. From this we conclude that it is the object's status as inexistent that causes satisfaction or jouissance. In the terms used in *The Project*, the *Nebenmensch*, or Thing, evades judgment; thinking encounters a limit, an impossibility beyond which it cannot go. On the other side of this limit there is nothing, no existing thing, nothing to think. On this side, however, there is not merely an experience of absolute impasse, of thought's negation. There is also affirmation in the form of satisfaction in an inexistent object, in an object that escapes the judgment of existence.

Deleuze attributes to Leibniz an argument that lights the path Freud will take:

> I *must* have a body because an obscure object lives in me . . . Leibniz's originality is tremendous. He is not saying that only the body explains what is obscure in the mind. To the contrary, the mind is obscure, the depths of the mind are dark, and this dark nature is what explains and requires a body.[42]

Mind encounters a limit, an obstacle, which is nothing more (or less) than an inexistent object, a darkness. Yet this obscure object does not merely check the powers of mind; it also incites an unshakeable conviction: there *must* be a body. Body does not impinge on mind and thereby obstruct it, nor for that matter does the mind collide with the body in its impenetrable density. Mind and body do not encounter one another, rather: mind encounters an obscure object that is neither purely internal nor purely external to it and it is this object which persuades mind that something other must exist.

Leibniz does not define this obscure object which disjoins/links *res cogitans* and *res extensa* as "satisfying" or as the object-cause of jouissance. It will be left to psychoanalysis to elaborate this dark spot in the mind in terms of libido or drive, as the frontier between lived experience and biological life. One can find, however, among earlier philosophers – namely the medieval followers of Avicenna, for whom the concept of just such a frontier or *barzahk* (in Arabic) played a major role in their thinking – a certain precedent for the direction in which Freud would develop the Leibnizian notion of a mental darkness. For those medieval philosophers the limit disjoined/linked the divine and the sensible worlds, but precisely because this limit passed not simply between these two worlds but also *through the sensible*, it (the limit or *barzahk*) was often conceived in terms of a dialectic of erotic love, as the disjoining/linking of lovers. And for Ibn 'Arabi, at least, the real object of love was considered to be not what was obtained, that is, not the beloved as such, but rather "something nonexistent . . . The object of loving adhesion in the moment when the lover has achieved union - is again something nonexistent, namely, the continuation and perpetuation of that union."[43] That is, what one loves in the other is "not a datum existing *in actu*," but not a mere nothing either.[44] Ibn 'Arabi brings this insight even closer to the Leibnizian dark spot by insisting that "Love [which] is closer to the lover than is his jugular vein," is so "excessive in its nearness that it acts . . . as a veil."[45]

In that compact space in which lovers, the sexes, *res cogitans* and *res extensa* "hold each other tight," what one adheres to is not something one obtains or grasps but that which escapes one's grasp inasmuch as it inexists, has not yet happened. However long it lasts, *real* love is enduring, it unites itself with a future that is not merely a receding horizon – if not now, perhaps tomorrow – but the future's proleptic event in the form of a surplus pleasure or (in Freud's phrase) an "incentive bonus" that promises *more* (*encore*) to come. This future, which arrives before it is actualized, is instigated by an encounter that is contingent, by a meeting with chance, with the unexpected. Here, clearly a distinction is registered between a future that is anticipated, awaited, but forever put off, a future always incompletely achieved – let us call it capitalism's future – and the amorous future, which overtakes us (and the chronological order of things) by surprise, as fore-pleasure. In his work on jokes, Freud seems to define the incentive bonus or fore-pleasure as a pleasure that slips through or hoodwinks the censor, but it makes more sense in terms of his theory to interpret fore-pleasure as an *affirmation* of the limit, or the hoodwink of the censor, a smuggled-in statement of the impossible: there is something I can neither know nor control. And yet I know this, for through the veils and dark spots, the obscurity in the depths of my mind, this truth *speaks*, for this obscurity has its roots not only in my mind but also in what is beyond it: the other, the body.

the return to two

The two of sexual difference must be thought in just this way. Not as two separate and opposed ones, not "that binary partition one most spontaneously thinks of [as] 'sexual difference,'"[46] a "pre-dual" sexuality, "more originary than the dyad" to which sexual difference is commonly reduced.[47] More originary than the dyad is the cut, the split, which is not a split into two "determinities [*Bestimmtheiten*]"[48] (that is: not two determinate ones), *nor* an intervention in an originary One. For, the One is not that which is split, but rather that which is formed from the splitting. Because it is thus formed, the One is paradoxical, a severed One, detached from some (surplus) part of itself from the start: $1 = a$.[49]

Derrida's argument in "*Geschlecht*: Sexual Difference, Ontological Difference" is that Heidegger chose the *neutral* term *das Dasein* for that form of being that places its own being in question rather than *man* (*Mensch*) not in order to disavow the ontological status of sexual difference but to distinguish it from the common understanding of it as a dyadic structure. This would make Heidegger's position parallel to that of Freud, who (we have already remarked) adamantly maintained (against feminist protests) that there was only one libido and it was male. To insist that being and libido are not divided into two *determinities* is not to say that either is *neutral* in the ordinary sense of the term but to say, on the contrary, that they are marked originally by sexual difference. Feminine sexuality, that dark continent of psychoanalytic thought, is not Leibniz's dark spot in the mind, rather: the latter is the mark, the result, of the former's withdrawal. The fact that the One is paradoxical, always more than itself, is coterminous with the fact that it is less than itself, that is: that something has been subtracted from it. Something always escapes the One. Thus, while Freud maintains that there is only one libido and it is male, this does not prevent him from puzzling over the question of that libido which cannot be counted as male and cannot be counted as one, not even another one because it is only thinkable as other.

The subject, insofar as he is subject of libido, is always singular; sexed, he is obliged to choose his own objects of satisfaction, carve out his own path. Yet, as I have noted, his singular or exceptional status does not negate but rather confirms or manifests the One, which exists only as these diverse exceptions. At the same time, however, there are some subjects whose singularity does *not* manifest the One and precisely insofar as they do not, insofar as they fall out of the One, they *do not consist*. Feminine sexuality, in short, cannot constitute women as a group.

notes

1 Teresa de Lauretis, *Technologies of Gender: Essays in Theory, Film, and Fiction* (Bloomington and Indianapolis: Indiana UP, 1987) x.

2 Shulamith Firestone, *The Dialectic of Sex* (New York: Bantam, 1972) 10–11.

3 For a thorough analysis of these questions see Gilbert Simondon, "The Genesis of the Individual" in *Incorporations*, eds. Jonathan Crary and Sanford Kwinter (New York: Zone, 1992) 297–319.

4 For more on this debate see, for example, Jorge Gracia, ed., *Individuation in Scholasticism: The Later Middle Ages and the Counter-Reformation* (Albany: State U of New York P, 1994). My own interest in the question of individuation was ignited by the defense of the medieval Islamic philosophers' position on this subject mounted by Henry Corbin; see his "Apophatic Theology as Antidote to Nihilism," trans. Roland Vegso, *Umbr(a): Semblance* (2007): 59–83. This essay was originally published as a chapter in Corbin's *Le Paradoxe du monotheism* (Paris: l'Herne, 1981).

Working from a different set of sources and questions, Mladen Dolar has begun a similar questioning of the two of sexual difference on the basis of a more sophisticated notion of the One in an excellent unpublished manuscript, "One Splits into Two." See also the fine work of our colleague Alenka Zupančič on the problematic nature of the two in *The Shortest Shadow: Nietzsche's Philosophy of the Two* (Cambridge, MA: MIT P, 2003).

5 Christian Jambet discusses the Islamic concept of the "unity of God" (*tawid*, in Arabic), which this formulation expresses, in his remarkable essay "The Stranger and Theophany," trans. Roland Vegso, *Umbr(a): The Dark God* (2005): 27–42. This essay was originally published in Jambet's *Le Cache et l'apparent* (Paris: l'Herne, 2003). Lacan – not coincidentally – employs the Islamic formula "There is no other God but God" in his reading of the "specimen dream" of psychoanalysis, the dream of Irma's injection, in order to drain the formula for trimethylamine (a product of the decomposition of sperm) of its sexual *substance* and reconstitute it as a empty signifier, a signifier that because it does not mean anything is able to indicate that excess in language which gives rise to sex. Jacques Lacan, *The Ego in Freud's Theory and in the Technique of Psychoanalysis*, ed. Jacques-Alain Miller; trans. Sylvana Tomaselli (New York: Norton, 1988) 158.

6 See Henry Corbin, *Alone with the Alone: Creative Imagination in the Sufism of Ibn 'Arabi* (Princeton: Princeton UP, 1969) 210. In Arabic the phrase "eternal haecceity" is *'ayn thabita*; it means "eternal 'essence' or quiddity," but no mere translation does the term – which requires considerable theoretical unpacking – justice.

7 Jacques Lacan, *Encore: On Feminine Sexuality, the Limits of Love and Knowledge: Book XX*, ed. Jacques-Alain Miller; trans. Bruce Fink (New York: Norton, 1998) 3.

8 For a lucid discussion of failed vs. real repetition, I recommend Alenka Zupančič, *The Odd One In: On Comedy* (Cambridge, MA: MIT P, 2008) 149–82.

9 Søren Kierkegaard, *Fear and Trembling/Repetition*, ed. and trans. Howard Hong and Edna Hong (Princeton: Princeton UP, 1983) 123.

10 Ibid. 131; Sigmund Freud, *Project for a Scientific Psychology* in *The Standard Edition of the Complete Psychological Works of Sigmund Freud* (SE), trans. James Strachey (London: Hogarth and Institute of Psycho-Analysis, 1966) 1: 356.

11 Gilles Deleuze, *Cinema 1: The Movement-Image*, trans. Hugh Tomlinson and Barbara Habberjam (Minneapolis: U of Minnesota P, 1986) 6.

12 Freud, "A Note upon the 'Mystic Writing-Pad'" in SE 19: 231.

13 Juliet Mitchell, "Introduction I" in *Feminine Sexuality: Jacques Lacan and the École freudienne*, eds. Juliet Mitchell and Jacqueline Rose (New York: Norton, 1982) 20.

14 Fethi Benslama uses the term "monotheistic" in relation to sex in the title of his interesting essay "Le Sexuel monotheiste et sa traduction scientifique," *Cliniques méditerranéennes* 73 (2006): 89–95, without, however, telling us what this adjective means in this context. I was happy to come upon this text in the midst of writing my argument here, since Benslama confirms my own regional tale of a Western retiring of the term *sexual difference* in favor of *gender* during the 1980s. As it turns out – Benslama recounts – the Arabo-Islamic world – once thought by Foucault, among others, to be the last bastion of an *ars poetica* against the *scientia sexualis* that steadily took over the West since the nineteenth century – became subject in the 1980s to this particular form of "Westoxification." At this point what had been the most common Arabic word for sex, *farj*, was rapidly replaced by the term *jins*, from the Latin *genus* or gender. And as *jins*, or gender, usurped the place of *farj*, the word that had for centuries been used for men *and* women, *farj* simultaneously became restricted in scope and began to designate the sexual organ of women *only*. At the same time, *jins*, which carried with it scientific, specifically

bio-medical, connotations absent from *farj*, narrowed the sense of sexual relations or affairs to the genital register.

15 Jacques Lacan, *The Other Side of Psychoanalysis*, Book XVII, ed. Jacques-Alain Miller; trans. Russell Grigg (New York: Norton, 2007) 144.

16 Ibid. 208.

17 Michel Foucault, *The History of Sexuality, Volume I: An Introduction*, trans. Robert Hurley (New York: Pantheon, 1978) 143.

18 Ibid.

19 Michel Foucault and Ludwig Binswanger, *Dream and Existence*, ed. Keith Hoeller (Atlantic Highlands, NJ: Humanities, 1993) 102.

20 Foucault, *History of Sexuality* 140–41.

21 Ibid. 157; Foucault also calls sex a "fictitious point" (156) and an "imaginary point" (155).

22 Ibid. 154.

23 Ibid.

24 Jacques Lacan, *Television: A Challenge to the Psychoanalytic Establishment*, ed. Joan Copjec; trans. Denis Hollier, Rosalind Krauss, Annette Michelson, and Jeffrey Mehlman (New York: Norton, 1990) 30.

25 Lacan, *Encore* 11.

26 Henry Corbin, "Prophetic Philosophy and the Metaphysics of Being" in *The Voyage and the Messenger: Iran and Philosophy* (Berkeley: North Atlantic, 1998) 208.

27 Freud, "New Introductory Lectures on Psychoanalysis. Femininity" in SE 22: 116; my emphasis.

28 Lacan, *Encore* 21.

29 Freud, "Femininity" 114.

30 Idem, "Instincts and their Vicissitudes" in SE 14: 122.

31 Idem, "Project for a Scientific Psychology" in SE 1: 331. This comparison between Descartes and Freud is made by Monique David-Ménard in "Sexual Alterity and the Alterity of the Real for Thought," *Angelaki* 8.2 (2003): 137–50.

32 Lacan, *Encore* 112.

33 Ibid. 23.

34 Sigmund Freud, *Civilization and its Discontents* in SE 21: 65.

35 Ibid. 66.

36 Ibid.

37 Gilles Deleuze, *The Fold: Leibniz and the Baroque*, trans. Tom Conley (Minneapolis: U of Minnesota P, 1993) 47.

38 Lacan, *Encore* 9.

39 Ibid. 10.

40 Ibid. 3.

41 Freud, "On the Universal Tendency to Debasement in the Sphere of Love" in SE 11: 18; my emphasis.

42 Deleuze, *The Fold* 85.

43 Corbin, *Alone with the Alone* 155.

44 Ibid. 154.

45 Ibid. 156.

46 Jacques Derrida, "*Geschlecht*: Sexual Difference, Ontological Difference" in *A Derrida Reader*, ed. Peggy Kamuf (New York: Columbia UP, 1991) 386.

47 Ibid. 387–88.

48 Ibid. 393.

49 Lacan, *Encore* 49: "In other words, there are three of them, but in reality, there are two plus *a*. This two plus *a*, from the standpoint of *a*, can be reduced, not to the two others, but to a One plus *a*." My entire essay may be summarized as an attempt to spell out the meaning of these brief sentences.

a. kiarina kordela

BEING OR SEX, AND DIFFERENCES

In memoriam Claude Lévi-Strauss

"Sexual Difference between Psychoanalysis and Vitalism"; the title of the present collection of essays is explicit: next to sexual difference, at stake is also a theoretical difference. I would like to begin with the latter.

I (a) theoretical difference: lacan – deleuze

The theoretical difference between psychoanalysis – by which I mean the Freudian–Lacanian line of thought – and vitalism – interest in which has been rekindled by Deleuze in the wake of biopolitics and other discussions of life and the body – is particularly pertinent at this historical juncture, as for some decades now we have been witnessing debates that seem to pit "Lacanianism" and "Deleuzianism" against one another. The underlying assumption in these debates, which I would like to recast as a question, is not only whether there are differences between the two theories – which evidently there are – but also whether the two theories are fundamentally incompatible, so that it would be a contradiction in terms to say: "I am a Lacanian-Deleuzian" or "Deleuzian-Lacanian."

As a bridge to the first of a couple of representative voices that justify this question, let us recall the following passage from Lacan's notorious seminar on "The Purloined Letter":

> [T]he subjects . . . more docile than sheep, model their being on the moment of the signifying chain which traverses them . . . [T]he displacement of the signifier determines the subjects in their acts, in their destiny, in their refusals, in their blindness, in their end and in their fate . . . [W]illingly or not, everything that might be considered the stuff of psychology, kit and caboodle, will follow the path of the signifier.[1]

An astute reader of this passage may react in the following way:

> All this may appear very "subversive," if measured by the standards of the classical ideological notion of "autonomous subject" – but isn't this very Lacanian mechanism at work in what we call the "postindustrial society of consumption"; i.e., isn't the so-called "postmodern subject" the passive ground traversed by links in the symbolic chain reacting to images which regulate his or her "psychology," "destiny," and "fate," unable to exert control over this mechanism?

Indeed, many have criticized Lacanian theory along these lines. But the actual lines that I just appropriated happen to belong to an eminent Lacanian scholar, and their intended target is not Lacan but Deleuze and other so-called "Neo-Spinozists," including Spinoza himself. I only had to replace the Spinozian/Deleuzian terms "affects" and "passions" (which are the ones used in the original passage) with equivalent Lacanian terms, such as the "signifier," the "symbolic chain," etc. So, in the original passage, authored by Slavoj Žižek, one actually reads:

> but isn't this very *Spinozist* mechanism at work in what we call the "postindustrial society of consumption"; i.e., isn't the so-called "postmodern subject" the passive ground traversed by partial *affective* links reacting to images which regulate his or her "*passions*," unable to exert control over this mechanism?[2]

What is the moral of my contrived substitution of referent? Given that both Lacan and Deleuze *have* subverted "the classical ideological notion" of "autonomous subject," and that this subversion, whether performed by Deleuze or Lacan, could indeed be seen as the "mechanism at work in what we call the 'postindustrial society of consumption,'" the question, then, becomes whether there is a decisive difference between performing the said subversion by invoking something called "affect" or "passion," as opposed to something called "signifier" or "symbolic chain." Addressing this question, in turn, presupposes knowing what each of these terms stands for.

But, so as not to proceed unilaterally, let us approach the issue also from the "other side," with an example from the Deleuzian-type of psychoanalytic criticism. Juxtaposing one of the most fundamental concepts of psychoanalysis, the unconscious, to the Deleuzian trademarked concept of rhizomatics, Elizabeth Grosz writes:

> Rhizomatics... names a decentered set of linkages between things, relations, processes, intensities, speed or slowness, flows: proliferations of surface connections. In this sense, rhizomatics is opposed to hermeneutics, psychoanalysis and semiotics, each of which seeks, in its different way, to link an object (a text, a subject, a sign) with a hidden depth or latency – sense, the unconscious, the signified.[3]

I find the coexistence of "hermeneutics" *and* "psychoanalysis," or of "a hidden depth or latency," "sense," and the "signified" *and* the "unconscious" in one and the same bundle of terms rather startling. How can psychoanalysis be associated with hermeneutics when, in Lacan's words, "the effect of interpretation is to isolate in the subject a kernel... of *non-sense*"?[4] How can the unconscious be associated with sense and the signified when Lacan says that "meaning survives only deprived of that part of *non-meaning* that is... what constitutes in the realization of the subject... the unconscious"?[5] And how can the unconscious be linked to "a hidden depth or latency" that, moreover, as the passage implies, would provide a center for the subject, when, again citing Lacan, the "unconscious... is transindividual" and "is not at the disposal of the subject in re-establishing the continuity of his conscious discourse"; when "the unconscious is '*discours de l'Autre*' (discourse of the Other)," and "the fact that the symbolic is located outside of man is the very notion of the unconscious"?[6]

Or, to take another central tenet of Deleuze's (and Guattari's) work, the machinic conceptualization of subjectivity – which for several Deleuzians constitutes not only a non-psychoanalytic but even an anti-psychoanalytic innovation – had not Lacan always persisted that the unconscious is precisely a "machine" or an "automaton"?[7]

Yet, as anyone who has read both Lacan and Deleuze knows, their theories do not simply reaffirm, let alone replicate, one another, even if by means of other terms. There remains an undeniable and irreducible difference. The question is, then: is their theoretical difference to be understood, as the aforementioned debates seem to indicate, as opposition, incompatibility, or even merely occasional mutual criticism, other points of concurrence notwithstanding? Or is it to be understood, not unlike sexual difference, as the abyss of an incommensurability that, the amplitude of other similarities notwithstanding, persists and is required for any amplifying affirmation of an other that, by definition, is and will remain an unassimilable, unknown other – the true shared point between the parties being the recognition of the indispensability of this radical otherness?

Taking into account the abundance of common ideas expressed by the two fellow thinkers, as well as, on the one hand, Lacan's unconditional devotion to Freudian psychoanalysis, and, on the other, Deleuze's persistent criticism thereof, whether in his solo work or in his collaborations with Félix Guattari, we have to infer that if their difference is not to be reduced to any of the available degrees of sheer discord, there must be something truly Other – a Third – that both thinkers are after, and which the one can approach only through his pledge to, the other only through his assault of, this target field, psychoanalysis.

We can find one first and most accurate description of the object of Deleuze's attack – one arguably preceding and informing his own descriptions of his criticisms – in Jean Laplanche and Jean-Bertrand Pontalis's account of the long, at once historical and logical, itinerary of Freud's thought toward structuralism.[8] We could say that what Deleuze detests is the obstacle that Freud devoted his life to overcome, namely: the epistemological and methodological "guiding principle of Freud's practice," particularly eminent in the late 1800s, that postulated "the search for chronology, going backwards into time towards the first real, verifiable elements," and which made him a "prisoner of a series of theoretical alternatives," such as "internal–external, imaginary–real," subject–object, and their corollaries.[9] The sole way to extricate thought from these alternatives and to establish, in Deleuze's words, that the "phenomena of the unconscious cannot be understood in the overly simple form of opposition or conflict," is to show that such "conflicts are the result of more subtle differential mechanisms (displacements and disguises)" that take place, in Laplanche and Pontalis's words, on the level of "a pre-subjective structure, beyond both the strict happening and the internal imagery."[10] It has to be shown that the "unconscious is differential and iterative . . . serial, problematic and questioning," and not, as a "Hegelian post-Kantianism" (which Freud had troubles leaving behind) would want it, "oppositional," which is only the principle of "propositions of consciousness" with their "answers–solutions" orientation (DR 108).

It is not difficult to discern already here that what is often perceived in Deleuze's work as a criticism, and even refutation, of psychoanalysis is actually a munificent endowment of his life-long painstaking philosophical creativity toward guarding psychoanalysis against its most insidious peril: the temptation of conceiving the subject as an individual agent of events in chronological time, rather than as a structural, non-anthropomorphic, machinic network of differential relations in a "pure and empty form of time" void of "all content," including any "temporal content" (DR 86, 110). It is a struggle against "the long history of the distortion of the dialectic, which culminates with Hegel and consists in substituting the labour of the negative for the play of difference and the differential" (DR 268), and which, as Lacan remarks about Hegel, remains within "the limit of anthropology" and "humanism."[11] It is the same establishment – "of pious destructions, such as those performed by psychoanalysis under the benevolent neutral eye of the analyst," the "Hegel-style destructions," which are nothing but "ways of conserving" – that both Deleuze and Lacan are attacking.[12] But in different ways. The latter by fanatically affronting it, not without an array of derogatorily intended terms that range from "ego-psychology" to the "dentist's" conception of the "*cogito*," while always, excommunicated or not, clinging on to a true psychoanalysis as a regulatory principle or potentiality.[13] The former by obsessively preventing the term psychoanalysis from tainting potentiality, in a compulsively renewed series of assaults that safeguarded for him an almost uninterrupted contact with it – and, all this, without omitting to demarcate, from time to time explicitly, the sole ally and exception to this otherwise universal conspiracy of Oedipalization, as in the caveat:

> As to those who refuse to be oedipalized . . . the psychoanalyst is there to call the asylum or

> the police for help. The police on our side! – never did psychoanalysis better display its taste for supporting the movement of social repression... The fact that there are some, around Lacan, who are developing another conception of psychoanalysis does not mean that we should take no notice of the dominant tone in the most respected associations.[14]

In short, the difference between Deleuze and Lacan converges in the stance that: "Psychoanalysis is like the Russian Revolution; we don't know when it started going bad";[15] the key being that this recognition alone says nothing as to whether one should never again appeal to the term "communism" or endeavor forever to redeem it by reconstituting its entire semantic field. What is more, nor does either choice make the one or the other a truer communist.

I(b) narcissism, death drive, lack

But of more relevance to the theoretical difference between Lacan and Deleuze, than such eponymic or allegoric references, is their conceptual interplay. One of the most appropriate concepts for exemplifying at once the intertwining relation and the mirage of mutual aversion between psychoanalysis and Deleuze's thought is another central Lacanian concept: lack. Deleuze's work is often characterized, in Judith Butler's words, as a philosophy of the "plenitude of life" that "disjoin[s] desire from negativity," whereas it has become practically a joke that "you can't take the lack out of Lacan."[16] But to speak of lack we must pass through two other central psychoanalytic concepts: narcissism and the death drive (both of which directly relate also to the issue of sexuality).

In *Difference and Repetition*, after the first two syntheses of time and his characteristic tour de force in ever-reinventing argumentative repetition, Deleuze eventually arrives at the "third synthesis of time" – as "the spatial phenomenon" of "the empty form of time... separated from its content" with its "already" and "not yet" – which is executed by the "narcissistic libido" (DR 110). If narcissism alone is capable of obtaining the third synthesis of time, this is so because, when the "order of before, during, and after marks the division of the narcissistic ego in time," the "whole of time is gathered in the image of the formidable action as this is simultaneously presented, forbidden and predicted by the superego" (DR 110). For in the repetition of narcissism,

> the narcissistic ego repeats once in the form of the before or lack, in the form of the *Id* (this action is too big for me); a second time in the form of an infinite becoming-equal appropriate to the *ego ideal*; a third time, in the form of the after which realizes the prediction of the *superego* (the id and the ego, the condition and the agent, will themselves be annihilated)! (DR 110–11)

It is due to this total annihilation of "the condition and the agent" that, on the one hand, time is evacuated of the "before" and the "after," as they become through the narcissistic repetition two different intensities: "too much" and "too little"; and, on the other, the future – the repetition in which condition and agent become "too little" as they are annihilated by the superego – is the "new, complete novelty" that "constitutes the autonomy of the product, the independence of the work – which leaves nothing intact of the default [past] or the becoming-equal [present]" (DR 90). As for the "present" itself – the moment of the "becoming-equal," between the "too much" (past) and the "too little" (future) – this "constitutes the totality of time to the extent that it draws together the... before and after," and "must be called a symbol by virtue of [precisely] the unequal parts which it subsumes and draws together" (DR 89–90). For, as Laplanche and Pontalis put it, the "symbol" or "the sign" is what "describes both the object and its absence,"[17] two unequal parts that, by being brought together in the sign, make it, in Deleuze's words, "the unequal in itself" (DR 90). In other words, the unique value of narcissism lies in the fact that, by introducing the sign, it animates the law (superego), which alone can confer the fatal blow on conditions and agent so that the "repetition of the future" enables the emergence of the "new, complete novelty" (DR 90).

Here, Deleuze formulates one of the most fundamental theses of psychoanalysis, namely, that the function of the law necessarily transcends its role of sustaining the status quo and is, therefore, simultaneously the very condition of the emergence of the new – a new that is, of course, always liable to renewed cycles of repetition. As Deleuze and Guattari put it, "territorialities" circumscribed by any law are always already destined to be "shot through with lines of flight testifying to the presence within them of movements of deterritorialization and reterritorialization."[18] Or, as Lacan puts it, "transgression . . . only takes place if it is supported by the oppositional principle, by the forms of the Law."[19] And, given that "the law . . . is revealed . . . as identical with an order of language"[20] it is only expectable that the very moment that "the passive ego becomes narcissistic, the activity must be *thought*" (DR 110).

As the fatal denouement of the narcissistic prediction with "its crushing unity and . . . irreversible series" indicates, it "is by means of the death instinct that [the narcissistic ego] is reflected in the ego ideal and has a presentiment of its end in the superego, as though in two fragments of the fractured I" (DR 111). The inevitable result of the fact that in narcissism "the commandment" of the law "appeared" is that "through it I came to desire death."[21] In narcissism, death and the amorous game meet, so that "Thanatos" is "by no means the complement or antagonist of Eros" (DR 111). Rather,

> it is all in the same movement that there is a reflux of Eros on to the ego [whereby it becomes narcissistic], that the ego takes upon itself the disguises and displacements which characterize the objects . . . and that the death instinct appears, indistinguishable from that pure form . . . of that narcissistic libido. (DR 113)

Thus, "the complementarity between the narcissistic libido and the death instinct defines the third synthesis of time," time as a pure and empty form (DR 113). Just as the death drive is "not . . . the exception to the [pleasure] principle but . . . its *foundation*," the drive is also the foundation of desire.[22] Deleuze references "need" as the blueprint for exemplifying this supplementary relation:

> A phenomenon such as need can be understood in terms of "lack", from the point of view of action and the active synthesis which it determines, but as extreme "satiety" or "fatigue" from the point of view of the passive synthesis by which it is conditioned. (DR 77)

Here, Deleuze distinguishes between two points of view and their corresponding syntheses of time, the active (first) and the passive (second), which relate to the central division of the Deleuzian universe throughout his oeuvre between the actual and the virtual. Each, "lack" or "satiety," is a partial view of one and the same phenomenon, "need," but they are not for that matter equal: it is satiety, the virtual view, that conditions need to experience itself on the level of the actual as a lack that requires action, and to determine itself toward an action. In the transition from the virtual to the actual – which we could call the "generative" or "synthetic" transition – the pure and empty form of time stretches itself into a diachrony with a specific directionality, at the same time that it is filled with contents of all sorts.[23] Conversely, in the transition from the actual to the virtual – which we could call "genetic" or "analytic" – "the line of time forms . . . an eternally decentred circle," the "straight-line labyrinth" or empty form of time that always leads to the future of the "eternal return" of "a world which has rid itself of . . . the condition . . . and the agent" (DR 111, 115).

Similarly, from the perspective of action, desire appears as a lack that produces time as a project, as in the Hegelian conceptualization of desire, which Alexandre Kojève formulates as follows:

> Man essentially creates and destroys in terms of the idea that he forms of the Future. And the idea of the Future appears in the real present in the form of a Desire . . . Action that arises from *this* Desire engenders History. Hence there is *Time* only where there is *History*.[24]

This "dialectical understanding applies only to historical reality – that is, to the reality created by

Work according to a Project."[25] But from the point of view of the virtual, desire is lack as the pure past, "a past which was never present," that is, "a past as the contemporary of the present" (DR 102–03) and which Jean-Paul Sartre had indelibly designated (for both Deleuze and Lacan) as the lacked totality. In Sartre's words:

> A lack presupposes a trinity: that which is missing or the "lacking," that which misses what is lacking or the "existing," and a totality which has been broken by the lacking and which would be restored by the synthesis of the "lacking" and the "existing" – this is the "lacked."[26]

The "lacked" restored is desire as excessive satiety. Hence, from the perspective of the virtual, it does not matter "whether this totality has been originally given and is now broken . . . or whether it has never yet been realized."[27] Rather, what matters is that this lacked totality function as "the before or lack," contemporaneous with the present, that engenders the repetition in which the condition and the agent of the action conditioned by it are crushed in the future.[28]

Given that, even on the actual level, "desire finds the principle of its difference from need in the virtual object" (DR 106), already for Hegel "human desire" is "a Desire directed toward another Desire," and not any actual thing.[29] But, the desire of the Other is like the gaze of the Other: always "not a seen gaze, but a gaze imagined by me in the field of the Other."[30] (For this reason, the "first upsurgings of desire" coincide with the "origin of fantasy."[31]) And this means that:

> Fantasy . . . is not the object of desire, but its setting. In fantasy the subject does not pursue the object or its sign: he appears caught up himself in the sequence of images . . . As a result, the subject, although always present in the fantasy, may be so in a desubjectivized form . . . in the very syntax of the sequence in question.[32]

This is why "the unconscious desires, and only desires," but only "as a questioning, problematising and searching force which operates in a different domain than that of [action-oriented] desire and satisfaction" or lack (DR 106). The action-oriented desire of historical time "goes . . . from one actual form to another," whereas unconscious desire "goes from the virtual to the actual, that is, from structure to its actualizations."[33] And it is the satiety of the (virtual) structure, with its lack, its always missing "empty square [la case vide]," that conditions all possible quests of the actualization.[34] After all,

> to become an animal . . . to participate in movement, to stake out the path of escape in all its positivity, to cross a threshold, to reach a continuum of intensities that are valuable only in themselves, to find a world of pure intensities where all forms come undone, as do all the significations, signifiers, and signifieds [– all this is worth doing only] to the benefit of an unformed matter of deterritorialized flux, of nonsignifying signs . . .[35]

. . . or, in Lacan's words, to the benefit of "irreducible, *non-sensical* – composed of non-meanings – signifying elements" that constitute "the unconscious," that is, the Third as the epitome of the animal, the rhizome, and any other stand-in for non-sensical signs or intensities.[36]

Once it is clear that lack is a condition of structurality on the virtual level, you can no more take the lack out of Deleuze than of Lacan. Nor can Deleuze's universe stand without the presupposition of the death drive. Were this to be removed, the virtual "upper level" of the "two floors" of the "world" would collapse, and the Hegelian dialectic would rightly claim for itself full reign over the entirety of existence.[37] The struggle against this possibility is for the two allies (to quote the title of an essay included in this volume) indeed a matter of life and death.

The theoretical difference between Lacan's and Deleuze's works may be such that what Joan Copjec writes somewhere in *Imagine There's No Woman* becomes impossible: "Lacan's characterization . . . is surprisingly close to Deleuze's"[38] – except as a phrase that, by means of inverted litotes, expresses that there is precisely no surprise. Difference means a problematic and questioning closeness – only consciousness is trapped in the simple form of opposition and conflict. Thus, we already have stated something about sexual difference too.

2 "what is sexual difference *now?*"

This question, instigating and introducing the present collection of essays, functions as its underlying motto. I have placed the emphasis on the "now" so as to let it lead us to another question. In the 1980s, elaborating on Martin Heidegger's assertion that "*each age has one issue to think through, and one only*," Luce Irigaray suggested that "sexual difference is probably *the* issue in our time which could be our 'salvation' if we thought it through."[39]

At first sight, Irigaray's by now quite famous thesis seems straightforward and unambiguous: all thought up until the twentieth century has never thematized, addressed, or even been concerned with the issue of sexual difference. For thousands of years everything has been thought and presented from a masculine perspective which, what is worse, does not even, in Elizabeth Grosz's words, "acknowledge its investments in" masculinity but rather presents itself as "universal" and "neutral," thereby effecting a "neutralization or neuterization" of femininity.[40] It is only in our historical era that sexual difference emerges as a theoretical and political issue, therefore sexual difference is the one and only issue of specifically our era. Irigaray herself corroborates this understanding of her statement by further asserting immediately after the aforecited passage: "whether I turn to philosophy, to science, or to religion, I find this underlying issue still cries out in vain for our attention." Our era alone has the potential to alter the course of history because, Irigaray continues, "sexual difference would constitute the horizon of worlds more fecund than any known to date."[41]

As a way of teasing out not a contradictory but a different meaning in Irigaray's thesis – that sexual difference is the issue of specifically our era – I will pursue a brief retrospection of (by now all) dead white men, first through Claude Lévi-Strauss, who will swing us as far back as the ancient myths and as close as Sigmund Freud, to focus then, in the next section, on Immanuel Kant.

In his classical essay "The Structural Study of Myth," Lévi-Strauss offers an analysis of several ancient myths from various cultures, which arrives at the conclusion that "mythical thought always progresses from the awareness of oppositions toward their resolution."[42] The function of myth is to resolve – by which he means to negotiate, not excluding the sense of covering up – an irresolvable initial opposition. This function, Lévi-Strauss argues, is universal, whether the myth in question is the "Zuni emergence myth," in all its variations, or the "American Ash-Boy," the "European Cinderella," or the "Oedipus myth."[43] The latter "has to do with the inability, for a culture which holds the belief that mankind is autochthonous" – i.e., born from the earth, and hence from One – "to find a satisfactory transition between this theory and the knowledge that human beings are actually born from the union of man and woman," that is, from Two. Although the problem obviously cannot be resolved," Lévi-Strauss continues, "the Oedipus myth provides a kind of logical tool which relates the original problem – born from one or born from two – to the derivative problem: born from different or born from same" – born from different blood families or born from incest.[44] Through the introduction of the concept of incest, the Oedipus myth succeeds in providing some kind of resolution to this original problem: One or Two. And when, more than two millennia later, Freudian psychoanalysis will be preoccupied with incestuous and Oedipal scenarios, they will still depict, as Lévi-Strauss argued, the same myth, just another version of it, no less true or original than Sophocles'. For, in Lévi-Strauss's own words:

> Although the Freudian problem has ceased to be that of autochthony *versus* bisexual reproduction, it is still the problem of understanding how *one* can be born from *two:* How is it that we do not have only one procreator, but a mother plus a father?[45]

The Lacanian psychoanalytic response to this question, at least since Copjec's inspired and inspiring reading of the Lacanian formulas of sexuation, will always already have been given by Kant – and it is a quite counterintuitive response: we come from two (a mother and a father) not because it takes two kinds of genitalia and other bodily organs to conceive and gestate a human embryo; rather, we humans come from two because, when eventually the infant-mammal-to-become-human enters the realm of language and thought, its reason fails in *two* ways. There *is* sexual difference, and it must be addressed in all kinds of discourses, from philosophy, art, and science to religion, "because," to quote Copjec's succinct formulation, far from being "singular" as "is naively assumed," reason's "failure" occurs "in one of two different ways" – what Kant calls the mathematic and the dynamic antinomies; and this, in turn, means that the "'universal' subject" of philosophy and abstract thought "is *necessarily* sexed."[46] Kant's *Critique of Pure Reason*, therefore, must be added to the extant versions of the Oedipus myth insofar as it, too, largely deals with the question of whether we come from One or Two, and further – a question that we must re-pose ourselves – whether One and/or Two already entail multiplicity, and whether such an entailed multiplicity would nevertheless remain structurally reducible to the One and Two.

The point of this invocation of Lévi-Strauss, Kant, and, through them, thousands of years of thought, from archaic life to modernity, is not to lead us to the false question:

> if all this long tradition has, since time immemorial, been doing nothing else but thinking about sexual difference, what's the point of regurgitating such a trite issue again, and, as a matter of fact, of doing so as if it were the unique issue of "our time" and the source of "our" salvation?

This is a false question not only because, evidently, this tradition has not been speaking *for* women or feminism. To re-raise Irigaray's question differently is, rather, to ask: is sexual difference the issue of our time – the issue of "now" – insofar as nobody has ever addressed it before, or is it *our* issue because precisely *all* history, whatever else it might appear to have been concerned with, has always been addressing this and only this issue? Is this question the "real" of human history, always, again and again, tirelessly and in multifarious ways, circumscribed through the eons? And is modernism, and even more so postmodernism, the era that has to return directly to the "original problem," the "initial opposition" of sexual difference, rather than replacing it with some other more amenable pair of terms? To be sure, if sexual difference *is* the real of history then it cannot ever be directly described; it is at best only circumscribed, even as we may think, as is common today, that we are talking directly about it. Yet these questions cannot but persist, and it is with them as background that I would like us now to approach sex and sexual difference.

3 sex and sexual difference

The preceding remarks, based on Lévi-Strauss and Kant, indicate that sexual difference – a difference that is external, between human beings – is a derivative of a primary internal difference, a fissure in the One owing to reason's internal failure, because of which the One is at once in excess of and less than itself, Two *and* not-all One. It is only because reason taken as a totality or One does not coincide with itself that it generates sexual difference.

Operating in the same post-phenomenological paradigm as Lacan and Deleuze, which does not admit a separation between being and appearance (representation), we are necessarily led by the above preliminary conclusion to proceed in a monistic (Spinozian) manner and to presuppose Being as One substance that, prior (in the logical sense of the word) to the failure, is asexual. As Lacan puts it in *Encore*, this complete and total (virtual) substance or "being is the jouissance of the body as such, that is, as asexual (*asexué*)."[47] Albeit asexual, Being *is* "jouissance," understood as "libido, *qua* pure life instinct . . . immortal . . . or irrepressible - indestructible life," and "whose characteristic is not to exist."[48] In other words, the One of asexual Being is not actual but, rather, an indestructible potentiality, the sheer irrepressible power of Being's self-actualization: it is pure Sex, not as in sexual reproduction, in which the product presupposes both a cause other than itself and sexual difference, but as the irrepressible power of Being to generate itself. Under the monistic principle of the One, "a substance cannot be produced by anything else . . .; therefore it will be the cause of itself,"[49] which means that Sex or "*being as such* is the power of making itself actual."[50] Hence, we can state from the outset that Sex is the drive – that is, the death drive which is one with the "pure life instinct" (Eros) – insofar as "the activity of the drive is concentrated in this *making oneself* (*se faire*)."[51] Sex is virtual (potentiality), while its actualization is actual and, as such, is no longer Sex. And because Being or Sex is the cause of itself, it is always more-and-less than itself and does not coincide with itself, so that it is forced to relate to itself and, thus, is self-referential. As such, Sex, the asexual One, or Being is not solid or material, but a *differential* substance.[52]

By contrast, as Lacan continues to say, "sexual jouissance" – not the jouissance qua Being but that of each of the two sexes – "is marked and dominated by the impossibility of establishing . . . anywhere in the enunciable . . . the One of the relation 'sexual relationship' [*la relation rapport sexuel*]."[53] Let's break down this statement. While, notoriously, "there's no such thing as sexual relationship [*rapport sexuel*],"[54] there is the "relation 'sexual relationship' [*la relation rapport sexuel*]," and this is the relation of "the One" of Being or Sex to itself – but this relation exists only virtually (i.e., to repeat, its "characteristic is not to exist" actually), and therefore cannot be enunciated by the actual sexed subject. In fact, it is for this very reason that "there's no such thing as sexual relationship"; for if the "relation 'sexual relationship' [*la relation rapport sexuel*]" could be communicated, could be "established . . . in the enunciable," then there would be sexual relationship (*rapport sexuel*). The primary impossibility of the One to coincide with itself, this internal split of the One because of which it relates to itself, and which is unrepresentable, is the cause of sexual difference: *human beings have different rapports* (*rapports*) *with this primary relation* (*relation*) *of the One to itself – and this is what we call sexual difference.*

In his *Theory of Religion*, published posthumously the same year as Lacan's *Encore* and his appearance on television in 1973, Georges Bataille turned to the oldest known tribal beliefs and rituals to arrive at the conclusion that the "essence" of "religion" lies in the "search for lost intimacy," a yearning to restore absolute "immanence," prior to the split between subject and object, prior to the Two.[55] Through a long process of historical transmutations, we arrive at secular science, whose essence, Bataille maintains, consists in relating to the One via the Two, the absolute split between subject and object. But, whether it is the One or the Two that dominates in specific historical moments and modes of thought, Bataille's trajectory reveals that all historical discourses or epistemes, from religion to science, are responses to, attempts to negotiate, our rapports with the fissure of the One. In other words, the difference between historical discourses, too, is sexual, while the fissure of the One, One's relation to itself, Sex, is the transhistorical impetus – *clinamen* – of humanity.[56] One way or the other, *one* (whether a human individual or an entire collective) must have some rapport with the relation of the One.

But whether in the religious or the scientific mode, whether longing for or exorcising it, the One elides us; and the reason is that it involves a *surplus*. "One," by definition, means no Other within it, yet, being its own cause, being more-and-less than itself, it encompasses within itself an Other. For example, I cannot be One because of the unconscious; only a thought or an idea can be unconscious (nothing else can be repressed) but only consciousness can have thoughts and ideas, therefore the unconscious *is* part of consciousness, and *yet* it is not. It is a radical Otherness that, however tightly embraced with consciousness, remains unknown. There is this surplus that is both inside and outside the One so that it cannot form a closed whole. This is so because any differential system – as are Being and representation – is always self-referential, with its cause as a surplus appendix to, and within, itself. As the paradox of self-referentiality teaches us, a differential system is never a closed system, never all. It is like the set of all sets, about which it cannot be decided whether it itself is a member of itself (the set of all sets) or outside it; it cannot be decided whether all or not all is part of itself. It is at once all and not all: it is an *all-not-all* set.[57]

It must be stressed that this all-not-all character of the One is a structural relation, that is, a relation that occurs outside of linear or diachronic time, in the pure and empty form of time, void of any content,[58] whose repetition amounts to the eternal return of the radical novelty which, as we shall see, is precisely sexual difference. The reason for this emphasis is to draw the reader's attention to a distinction generally overlooked in Lacanian theories.

To approach this distinction, I now turn to Copjec who has also argued that "being in general . . . resists being assembled into a whole." Why does Copjec need to add "in general" after "being"? Because she has arrived at this conclusion by equating Being with the not-all (*pas-tout*) of the female sex. Having done so, she proceeds to argue that "the famous formulation of a feminine 'not-all' . . . is fundamentally an answer not just to the question of feminine being, but to being as such."[59] But by asserting that feminine being is being as such, don't we replicate that mechanism, unanimously criticized by Copjec and other feminists, in which "binary oppositions, while purporting to oppose disjunct terms, tend in fact to negate the negation of one of these terms, which is thereby rendered susceptible to sublation by the other term"?[60] As Copjec's use of the term "sublation" (*Aufhebung*) indicates, and as Žižek has amply made clear, this mechanism is central to the Hegelian dialectic and exemplary of phallocratic thought. In a constellation of a universal genus – Being – one of its two particular species (feminine and masculine) – the feminine – appropriates the position of the genus; in this case, "the paradoxical element, the specific point where the universal *genus falls on itself* among its particular species, is [the] very phallic element."[61] To maintain that feminine being – and, for that matter, either sexed being – is Being as such is a phallic gesture.

If we are so easily led to mistake sexed being for Being as such it is because it is easy to conflate two distinct structures, the all-not-all and the two sides of the Kantian antinomies. To make clear the distinction between Being or Sex and the antinomies of sexual difference, we have to go once more through the Kantian antinomies, the Copjecian matrix of Lacan's sexual difference.

As we shall presently see, in formulating his two antinomies Kant is in fact grappling with the all-not-all character of Being, the paradox of set theory, albeit at a remove. By this, I mean that the paradox of set theory lurks underneath Kant's thought too – but its murmur is heard not only through the mathematic/female antinomy but also through the dynamic/male. For the primary thing of which Kant is conscious is that Being (thing-in-itself) is self-referential, that is, it cannot form a totality, infinite or finite. This is why he has to take as his point of departure the central premise of the pre-phenomenological paradigm (which he represents) that separates Being from appearance, and to posit from the outset that, if anything can form a whole, be it finite or infinite, this is only appearances; by contrast, the "world" qua thing-in-itself (i.e., not as a appearance) "exists neither as *an in itself infinite* whole, nor as *an in itself finite* whole."[62] Only what is within time and space is appearance and can (possibly) form a whole; what is outside time and space is thing-in-itself and cannot (in any case) form a whole. The set underlying and motivating both antinomies in Kant is the world of appearances within time and space. The question therefore for both antinomies is: *Can appearances form a whole*, be it infinite or finite?

The dynamic antinomy – whose concern is the whole of causality, that is, the causal relations of everything in the world, and whose specific question is accordingly: is everything in the world subject to natural necessity (read: causal determinism) or is there also freedom? – concludes that in the realm of "appearance every effect is an event . . . that happens in time," and so does its cause, and the cause of its cause, and so on, "hence, natural necessity must be the condition according to which [all] causes [within the realm of appearances] are determined." "If freedom," Kant continues, is to be asserted, then it must be an unconditional cause, a cause that is not the effect of some other cause, that is, "it must not stand under time-determinations of its state," and hence, "it must be considered as a thing-in-itself, while only its effects would be appearances."[63]

And the dynamic antinomy resolves the contradiction in the following way:

> [N]atural necessity . . . attach[es] to all connections of cause and effect in the sensuous world [appearances], though, on the other hand, freedom can be granted to the cause which is itself not an appearance . . . Nature and freedom therefore can without contradiction be attached to the very same thing, but in different relations [*Beziehung*] – on one side as an appearance, on the other side as a thing in itself.[64]

Thus, the set of all appearances can form a closed set, about which it can be said that it is in its entirety subject to natural necessity, but only insofar as everything belonging to the set of appearances is also, in a different relation (*Beziehung*), a thing-in-itself outside of the set. By positing an exception (freedom), the male sex forms a closed whole (law), yet everything under the law can, in another relation, be free. Through the dynamic antinomy and its closed set transpires the paradox of the set theory, as everything can be inside or outside the set, depending on the given relation.[65]

On the other hand, the mathematic antinomy – which is concerned with the spatio-temporal limits of the world, and, hence, asks whether the world in itself is limited or unlimited in time and space – bears at first sight an air of absurdity. For, since we can speak of finitude or infinity only about things that *are* in time and space (i.e., appearances), not the world as a thing-in-itself, we cannot even ask whether the world in itself is finite or infinite. We can know nothing about the world-in-itself, in fact, strictly speaking, we cannot even know whether there is a world-in-itself beyond appearance. We must stop, therefore, at this "*indefinite judgment . . .* without," as Copjec puts it "pronouncing beyond this on the existence of the world."[66] All we *can* know is "that however far we may have come in the series of empirical conditions" through which we endeavor to identify the limits of appearances, "we should never assume an absolute boundary, but rather we should subordinate every appearance as conditioned to another as its condition, and . . . this is a *regression in indefinitum*," an indefinite regress – something which Kant emphatically distinguishes from "the regress *in infinitum*," as the former "determines no magnitude in the object" (the world-in-itself) but the magnitude of our "empirical regress."[67]

Thus, the universe of Woman does not form an all (is not-all) because there can always be another condition in the indefinite regress of the series of the conditioned, or in Copjec's more linguistically inflected version, because "there will always be another signifier" in the signifying chain "to determine retroactively the meaning of all [signifiers] that have come before."[68] Here too, however, as in the case of the dynamic antinomy, the paradox of set theory shines through this indefinite regress in time and space: for the set of appearances, albeit indefinite, could still form a whole if it could only answer the question as to whether there is or is not a world-in-itself outside it; if we could only know whether what we refer to when we speak of the "world-in-itself" really exists or is yet another of our appearances (signifiers); if we could only know whether the *"world-in-itself" is or is not a member of the set of appearances.*[69]

The series of signifiers or conditions regresses indefinitely *in time and space*, and this is why Woman is not-all – unlike Being or the One which is all-not-all, more-and-less than itself, because part of itself is both inside it and outside it, within the logical *synchrony* of the empty form of time without any content (temporal or spatial), on the virtual level of the death drive. Let us return to Copjec's words once more, placing the appropriate emphasis: "there will always be another signifier" in the signifying chain "to determine retroactively the meaning of all [signifiers] that have come *before*."[70] No sooner has sexual difference emerged than the "before" and the "after," time, and hence mortality, follow. Indeed, Lacan stresses emphatically that what the subject losses through the introduction of sexual difference is immortality, or inversely, that sexual difference introduces diachrony and mortality; to cite again Lacan's relevant passage in its more extensive form:

> what the sexed being loses in sexuality . . . is . . . immortal. Because it survives any

> division . . . the libido, *qua* pure instinct of life, that is, immortal life, or irrepressible life, life that has need of no organ . . . indestructible life . . . is precisely what is subtracted from the living being by virtue of the fact that it is submitted to the cycle of sexed reproduction.[71]

The introduction of sexual difference represents the transition from the virtual realm of the death drive qua pure and indestructible life instinct or Eros – which, "having renounced all matter . . . corresponds to . . . the empty form of time," and "refers to the state of free differences when they are no longer subject to the form imposed upon them by an I or ego [and] any identity whatsoever" – to the actual realm, in which death "is personal, concerning the I or the ego, something which I can confront in a struggle or meet at a limit, or in any case encounter in a present which causes everything to pass" (DR 112–13). In his depiction of the "two aspects" or faces of death – the impersonal death drive and the personal death – inspired by Maurice Blanchot, Deleuze concludes: "There is always a 'one dies' more profound than 'I die'" (DR 112–13).

I recapitulate, first, by stressing the difference between the all-not-all of Being and the female not-all: in set theory the set of all sets is all-not-all not because we perpetually encounter yet another set but because it cannot be decided whether it itself (the set of all sets) is or is not included as a member of itself. Examined from the perspective of set theory, the obstacle preventing a set from forming a totality is not the indefinite regression in the diachronic series of its elements but the *self-referentiality of its synchronic totality*. Unlike sexual difference, which introduces time and, hence, (actual or personal) death, the One substance, Sex (Eros), which coincides with the death drive, is not in time.

Second, to recapitulate by addressing both sexual rapports with the self-relation of the One, the dynamic antinomy or male sex negotiates the inherent contradiction of the One by assuming that everything is inside the One – defined as the field of law (the symbolic order) – only insofar as, in another aspect, everything is outside: not subject to any law but free. The mathematic antinomy or female sex deals with the same problem by raising the unanswerable question as to whether the One – defined as the field of appearances (again, the symbolic order) – includes everything within itself or whether there is a world outside appearances, a world in-itself. If the male and female subjects stopped here, diachronic time would not yet be introduced, and each sex in its own way would have arrived at the ethical state – a state which is outside, or a break in, time, and which, like Being, is structured as all-not-all.[72] But both men and women have to live also in time, which is to say, they have to produce it: man by creating the times at which he is free and those at which he is subject to the law; woman by creating the before and the after in which experience retroactively constitutes itself. In both cases, this production is the generative moment at which (empty) time moves from the virtual to the actual, thereby filling itself with content, temporal and otherwise. Thus, the two sexes emerge, each with its own actual time, out of the empty form of time of Sex qua Being. And because no representation enunciated from within their actual times can ever represent Sex or Being (the Other object of their rapport), sexual difference stands for the failure of representation and is, therefore, real.

4 sex and biopolitics

To approach biopolitics we must begin with the question regarding the difference between nature and culture – a question that amounts to the incest prohibition. In *The Elementary Structures of Kinship*, Lévi-Strauss superseded the dualism between nature and culture by arguing that the incest prohibition

> is in origin neither purely cultural nor purely natural . . . It is the fundamental step because of which, by which, but above all in which, the transition from nature to culture is accomplished. In one sense, it belongs to nature, for it is a general condition of culture . . . However, in another sense, it is already culture, exercising and imposing its rule on phenomena which initially are not subject to it.[73]

Paraphrasing Lévi-Strauss's point, the Japanese philosopher and literary critic Kojin Karatani writes:

> If we do not consider culture as an a priori given, it can only be deduced from nature. But culture does not emerge from nature. Lévi-Strauss solved this by turning to "prohibition"... The prohibition of incest is "what man makes," but it is not made by man, because it is this prohibition itself that makes man into man.[74]

To our knowledge, the first social law ever imposed is the one that prohibits the specific sexual practice of incest – *the self-referentiality of the One on the level of blood.* Self-referentiality must be prohibited on some level for society to be possible. The "prohibition of self-referentiality," as Karatani remarks, is what "[i]n psychoanalysis... is called castration," the necessary *pass* into the symbolic order (and castration is not repression – what is castrated does not return).[75]

With the incest prohibition – which evidently has been operative since time immemorial, already administering the bodies and sexual practices of tribal societies – we are already in the realm of biopolitics. The prohibition of incest and generally the laws regulating endogamy, exogamy, and all other complex configurations of sexual circulation in tribes are the primal forms of law. In fact, one is tempted to say that if there were no sexual practices, there would be no law either. In any case, for us sexual human beings, sexuality and the law are indissolubly intertwined, something which is testified both by psychoanalysis's positioning of castration at the basis of subjective and social constitution and Deleuze's convergence of the narcissistic libido with the superego. By the same token, politics has always been biopolitics, the attempt to administer and regulate the body and Sex or Being – under the condition that we understand: (a) that Sex or Being is the One's relation to itself (as a consequence of which there is sexual difference – after all, if the One were a non-relational solid thing who would ever even care to engage in sexual practices?); and (b) that biopolitics has nothing to do either with the repression of sexuality – one's rapport with the One's self-referentiality cannot be repressed, for repression applies only to signifiers, not to rapports or relations – or with its discursive production (as in Foucault's inversely symmetrical criticism of the "repression hypothesis").[76]

Not only the reign of the sovereign but already that of the tribe chief has always been imbued with biopolitics, insofar as politics (the symbolic order) has always been dependent upon the prohibition of the self-referential relation of the One. *Yet, even as biopolitics is, therefore, transhistorical, every specific historical era always enacts its own mutation of this transhistorical social phenomenon.* There is, therefore, a historical manifestation of biopolitics that pertains specifically to secular capitalist modernity – and which will be our last topic here.

In *Grundrisse,* Marx writes that labor-power is "the use-value which the worker has to offer to the capitalist [and which] is not materialized in a product, does not exist apart from him at all, thus exists not really, but only in potentiality."[77] As opposed to slave or serf labor, the advent of the "free laborer" of capitalism entails that what is bought and sold between capitalist and laborer is labor-power as a sheer potentiality with no material existence in itself. Labor-power, then, as Paolo Virno points out, "is not separable from the *living person* of the seller," the "living body of the worker," which "is the substratum of that labor-power which, in itself, has no existence." The body, understood as "pure and simple *bios*... acquires a specific importance in as much as it is the tabernacle of... mere potential"; therefore, Virno concludes, it is "for this reason, and this reason alone, [that] it is legitimate to talk about 'bio-politics.'"[78] Here, Virno implies that body or bios (Being) is a *material* substratum or tabernacle of potential (immaterial) labor-power. As we have seen, however, it is the body itself, bios or Being, that has *always* been potential or immaterial, that is, *a relation* – which is why it has always been legitimate to talk about biopolitics. For, to repeat, biopolitics is precisely the administration of the self-referential relation of the One (Sex or Bios) to itself.

Yet, even as it has always been legitimate to talk about biopolitics, we *did not.* Why? Because, if something really changed with the advent of capitalism and its commodification of labor-power, this is rather, as Virno rightly remarks, the *catholic commodification of the potential, or the relational and differential* – which amounts to the direct *commodification of both the sexed body and Being.* The actual shift consists, in Virno's own words, in that "primary fact – both historical and philosophical – . . . of the commerce of the potential as potential," that is, the commerce of Sex.[79] The capitalist commodification of the potential or differential – in one word, of substance – indicates, as Lacan has put it, what the "starting point" of "capitalism" will always already have been, namely: "getting rid of sex."[80] The advent of free labor-power in the market entails the attempt to inscribe within (economic) representation that which persists only as long as it cannot be established in the enunciable – Sex.

Moreover, capitalist technologies eventually transformed human life into what since Guy Debord we have called the "society of the spectacle." Next to labor-power, images are the foremost commodities. Images, in turn, presuppose gazes.[81] What is the effect of the specularization of production on biopolitics? What is the relation between the gaze and the body or Sex? We all know that "[w]hat we call the body," as Lacan says, is "what lies under the habit," which, however, we easily mistake for a "substance . . . supposedly object-like (*objectal*)" forgetting that "in fact [it is] a remainder in desire, namely, its cause . . . what I call *objet a*" – whose manifestation in the field of vision or "scopic level" is the "gaze."[82] In the specularized society of secular capitalist modernity, the body *is* the gaze. It was again Sartre who had argued (prior to both Lacan and Maurice Merleau-Ponty) that this is the case. In his words from *Being and Nothingness*: "on the top of that hill which I call a 'good viewpoint,' I take a point of view at the very instant when I look at the valley, and *this point of view on the point of view*" – the point of view that could see my point of view looking at the valley – "is my body"; that is, my body is "the point of view on which I can *no longer* take a point of view." For it is "impossible to withdraw in order to 'give oneself plenty of room' and to constitute a new point of view on the point of view"; only an Other could have plenty of room.[83] Only an Other could take this point of view that sees my point of view while I am looking at something. *I am One that requires to be Two* in order to see myself seeing. In secular capitalist modernity it is the gaze that comes to incarnate the impossibility because of which my Oneness is fissured, and, thus, to embody the paradox of the self-referentiality of Being.

It follows, Sartre writes, that "I cannot take a point of view on my body *without a reference to infinity*" – which Lacan reformulates as: "*where there is being, infinity is required.*"[84] *Where there is Being or Sex* (self-referentiality) *both Two and infinity are required.* To say there is Sex and then there are two different sexes by definition means that there are not just (echoing Deleuze and Guattari, Grosz, and others) a "thousand tiny sexes" but a theoretically infinite, practically indefinite, number of teeny-bitty sexes – *but none of them could ever cut its umbilical cord from the initial duality of the Two, and from their own progenitor, the fissured One, Sex itself.*

Things *do* change, immensely, and, for that matter, a little. In the pre-capitalist kingdom, it was religion that aimed at getting rid of Sex – the *fissured* immanence of the One – by devoting itself to the complete plenitude of *total or absolute* immanence, a non-self-relational immanence. And, in the same move, it was religion itself that assumed the fissure of Sex in the first place by externalizing it as the dualism between the sacred and the profane in all its permutations, up to the dualism between mind and body, and subject and object. Always the homeopathic logic: the wound can be healed only by the spear that caused it. In the empire of capital, every economic aspect assumes the fissure of Sex: commodities that are both use-values and exchange-values; money that is both currency and a commodity itself; labor-power that is both potential and a commodity; and, with informatized capitalism, means of production that are material and immaterial. And, again, in the same move, capitalism itself aims at getting rid of Sex – and, for that matter, of religion, as its competitor – by proposing the indefinite plurality of its products, material and not, as the new arena in which total immanence can be obtained. *E pluribus unum*, as the Great Seal of the United States of America has it.

Resisting the reduction of everything to pluralism is necessary if we want to continue to have sexual difference now. Otherwise we are left in a world that claims not only that only consumerism can heal the individual's wounds but also that only global capitalism can heal all collective social wounds. In short, secular thought is a religion that, like all religions, strives for immanence; yet it differs from other religions in acknowledging that immanence, far from being absolute, is itself fissured; secular thought becomes secularist, pro-capitalist, and exploitative when it reduces fissure to some unqualified and uncaused multiplicity, whether in its neutrally asexual or harmoniously poly-sexual variant.

There has never been any manifestation of social life, artistic, philosophical, historical, religious, economic, or political, that does not address the immanence and fissure of Sex. Indeed, "*each age has one issue to think through, and one only*": the issue of Sex and sexual difference.

notes

1 Jacques Lacan, "Seminar on 'The Purloined Latter,'" trans. Jeffrey Mehlman, in *The Purloined Poe: Lacan, Derrida and Psychoanalytic Reading*, eds. John P. Muller and William J. Richardson (Baltimore: Johns Hopkins UP, 1988) 44.

2 Slavoj Žižek, *Tarrying with the Negative: Kant, Hegel, and the Critique of Ideology* (Durham: Duke UP, 1993) 218; emphasis mine.

3 Elizabeth Grosz, "A Thousand Tiny Sexes: Feminism and Rhizomatics," *Topoi* 12 (1993) 174.

4 Jacques Lacan, *The Four Fundamental Concepts of Psychoanalysis*, ed. Jacques-Alain Miller; trans. Alan Sheridan (New York: Norton, 1981) 250.

5 Ibid. 211.

6 Jacques Lacan, *Écrits: A Selection*, trans. Alan Sheridan (New York: Norton, 1977) 49, 312; idem, *Écrits: The First Complete Edition in English*, trans. Bruce Fink, with Héloïse Fink and Russell Grigg (New York: Norton, 2002) 392.

7 Idem, *Book II: The Ego in Freud's Theory and in the Technique of Psychoanalysis 1954–1955*, ed. Jacques-Alain Miller; trans. Sylvana Tomaselli (New York: Norton, 1991) 73; idem, *Four Fundamental Concepts* 52 *passim*.

8 The article by Laplanche and Pontalis to which I refer here is referenced recurrently by Deleuze, and has evidently had a major impact on his thought: Jean Laplanche and Jean-Bertrand Pontalis, "Fantasy and the Origins of Sexuality" in *Unconscious Fantasy*, ed. Riccardo Steiner (London: Karnac, 2003); originally "Fantasme originaire, fantasmes des origins, origine du fantasme," *Les Temps Modernes* 19.215 (1964). As Edward Kazarian remarks: "it seems that some of the 'credit'... should be given to Laplanche and Pontalis" for the major shift that occurred in Deleuze's thought between 1961 and "some time after this, but before 1967" – notably, his reevaluation of Freud, the concomitant abandonment of "his forceful endorsement of Jung," and the effects of his exposure to the "direction of 'structure' in a sense that Deleuze will adopt." Edward P. Kazarian, "The Revolutionary Unconscious: Deleuze and Masoch," *SubStance 122* 39.2 (2010) 106 n. 17.

9 Laplanche and Pontalis, "Fantasy and the Origins of Sexuality" 119.

10 Gilles Deleuze, *Difference and Repetition*, trans. Paul Patton (New York: Columbia UP, 1994) 106, henceforth cited as DR; Laplanche and Pontalis, "Fantasy and the Origins of Sexuality" 118.

11 Jacques Lacan, *Book II: The Ego in Freud's Theory* 72.

12 Gilles Deleuze and Félix Guattari, *Anti-Oedipus: Capitalism and Schizophrenia*, trans. Robert Hurley, Mark Seem, and Helen R. Lane (Minneapolis: U of Minnesota P, 1983) 311.

13 Lacan, *The Ego in Freud's Theory* 3 *passim*.

14 Deleuze and Guattari, *Anti-Oedipus* 81. While on the *ad hominem* aspect of the theoretical difference between Lacan and Deleuze, one of the latter's own anecdotes about their transferential relation seems in place. In an interview – worth reading in its entirety – shortly before his death,

Deleuze was asked by Didier Eribon about his relationship to Lacan, and Deleuze responded: "Lacan noticed me when he devoted a session of his seminar to my book on Sacher-Masoch [1967]. I was told . . . that he had devoted more than an hour to my book," and a "few months . . . after the appearance of *Anti-Oedipus* [1972]," about which Deleuze thought that Lacan would take "it badly[,] he summoned me – there's no other word for it. He wanted to see me. And so I went," to be told by Lacan: "C'est quelqu'un comme vous qu'il me faut" (What I need is someone like you). "Le 'Je me souviens' de Gilles Deleuze," interview with Deleuze by Didier Eribon, *Le Nouvel Observateur* 1619 (16–22 Nov. 1995) 50–51. Excerpts from this interview are cited, with interesting commentary on the further triangulation Lacan–Deleuze–Žižek, in Daniel W. Smith, "The Inverse Side of the Structure: Žižek on Deleuze on Lacan," *Criticism* 46.4 (2004) 635–38.

15 Deleuze and Guattari, *Anti-Oedipus* 55.

16 Judith Butler, *Subjects of Desire: Hegelian Reflections in Twentieth-Century France* (New York: Columbia UP, 1999) 212; Bruce Fink, "Knowledge and Jouissance" in *Reading Seminar XX: Lacan's Major Work on Love, Knowledge, and Feminine Sexuality*, eds. Suzanne Barnard and Bruce Fink (Albany: State U of New York P, 2002) 34. Bruce Fink attributes this line to Shelly Silver. For further commentary on Butler's thesis see Dorothea Olkowski, *Deleuze and the Ruins of Representation* (Berkeley: U of California P, 1999).

17 Laplanche and Pontalis, "Fantasy and the Origins of Sexuality" 131.

18 Gilles Deleuze and Félix Guattari, *A Thousand Plateaus: Capitalism and Schizophrenia*, trans. Brian Massumi (Minneapolis: U of Minnesota P, 1987) 55.

19 Jacques Lacan, *Book VII. The Ethics of Psychoanalysis, 1959–1960*, ed. Jacques-Alain Miller; trans. Dennis Porter (New York: Norton, 1992) 177.

20 *Écrits*, trans. Sheridan 66.

21 Lacan, *Ethics of Psychoanalysis* 83.

22 Gilles Deleuze, *Masochism: Coldness and Cruelty* (New York: Zone, 1989) 113.

23 "Time" is "genesis," "from the virtual to the actual" (DR 180).

24 Alexandre Kojève, *Introduction to the Reading of Hegel: Lectures on the Phenomenology of Spirit*, assembled by Raymond Queneau; ed. Allan Bloom; trans. James H. Nichols Jr (Ithaca, NY: Cornell UP, 1980) 139.

25 Ibid. 146.

26 Jean-Paul Sartre, *Being and Nothingness: An Essay on Phenomenological Ontology*, trans. Hazel E. Barnes (New York: Washington Square, 1992) 135. This reference to Sartre's thought is not to deny in general its considerable Hegelian tint but rather acknowledges that Sartrean "lack," far from being reducible to a kind of dialectical negativity, bespeaks, to borrow from Deleuze's own homage to Sartre, a "faithful" commitment "ever and always to the friend Pierre-who-is-never-there," to the "peculiar destiny to circulate pure air . . . even if this pure air, the air of absence, is difficult to breath." Gilles Deleuze, "He was my Teacher" in *Desert Islands and Other Texts 1953–1974*, ed. David Lapoujade (Los Angeles: Semiotext(e), 2004) 80.

27 Sartre, *Being and Nothingness* 138.

28 Thus, as in a repetition of World War I's primary murder of Freud's "wishful thinking" and consummation of the advent of the death drive in the world, World War II confers, in Deleuze's epic cinematographic reconstitution of time, the fatal blow because of which cinema becomes capable of the act that was to end action: the movement-image is crushed, and the novelty of the time-image emerges to express the empty form of time, "time . . . no longer subordinated to movement" (DR 89). See also Deleuze, *Cinema 1*, and *Cinema 2: The Time-Image*, trans. Hugh Tomlinson and Robert Galeta (Minneapolis: U of Minnesota P, 1995). Zsuzsa Baross's reading of Deleuze via Blanchot also argues that death produces the new as the time-image; "Lessons to Live: Deleuze," *Deleuze Studies* 3.2 (2009): 162–84. My gratitude to Cesare Casarino for referring me to this work.

29 Kojève, *Introduction to the Reading of Hegel* 139.

30 Lacan, *Four Fundamental Concepts of Psychoanalysis* 84.

31 Laplanche and Pontalis, "Fantasy and the Origins of Sexuality" 131.

32 Ibid. 133.

33 Gilles Deleuze, "How do we Recognize Structuralism?," trans. Melissa McMahon and Charles J. Stivale, in *Desert Islands* 180.

34 Ibid. 184.

35 Gilles Deleuze and Félix Guattari, *Kafka: Toward a Minor Literature*, trans. Dana Polan (Minneapolis: U of Minnesota P, 1986) 13.

36 Lacan, *Four Fundamental Concepts of Psychoanalysis* 250.

37 Gilles Deleuze, *The Fold: Leibniz and the Baroque*, trans. Tom Conley (Minneapolis: U of Minnesota P, 1994) 102.

38 Joan Copjec, *Imagine There's No Woman: Ethics and Sublimation* (Cambridge, MA: MIT P, 1994) 225.

39 Luce Irigaray, *An Ethics of Sexual Difference*, trans. Carolyn Burke and Gillian C. Gill (Ithaca, NY: Cornell UP, 1993) 5.

40 Grosz, "A Thousand Tiny Sexes" 167.

41 Irigaray, *An Ethics of Sexual Difference* 5.

42 Claude Lévi-Strauss, *Structural Anthropology*, trans. Claire Jacobson and Brooke G. Schoepf (New York: Basic, 1963) 224.

43 Ibid. 213–31.

44 Ibid. 216.

45 Ibid. 217.

46 Joan Copjec, *Read my Desire: Lacan against the Historicists* (Cambridge, MA: MIT P, 1994) 212–13.

47 Jacques Lacan, *Book XX. Encore, 1972–1973: On Feminine Sexuality; The Limits of Love and Knowledge*, ed. Jacques-Alain Miller; trans. Bruce Fink (New York: Norton, 1998) 6.

48 Lacan, *Four Fundamental Concepts of Psychoanalysis* 197.

49 Baruch/Benedict de Spinoza, *Ethics* in *The Collected Works of Spinoza*, ed. and trans. Edwin Curley (Princeton: Princeton UP, 1985) 412; Part I, prop. 7, dem.

50 Beth Lord, *Spinoza's Ethics* (Bloomington: Indiana UP, 2010) 21.

51 Lacan, *Four Fundamental Concepts of Psychoanalysis* 197, 195.

52 Elsewhere I have argued more extensively that in secular ontology Being can be conceived only as differential; A. Kiarina Kordela, *$urplus: Spinoza, Lacan* (Albany: State U of New York P, 2007).

53 Lacan, *Encore* 6–7.

54 Ibid. 34.

55 Georges Bataille, *Theory of Religion*, trans. Robert Hurley (New York: Zone, 1989) 56–57.

56 I am referring to the atomistic (represented particularly by the line Democritus, Epicurus, Lucretius) concept of *clinamen*, a minimal indeterminacy in the motions of atoms, an unpredictable swerve, without which we would be "automata," that is, entirely predictable – assuming that we and all nature would have existed in the first place, something which, according to the atomists, would have been impossible because in that case no collision between atoms would take place and no nature out of atoms would be created. Lacan explicitly bases his argument in the section "Tuché and Automaton," in Seminar XI, on this opposition between "automaton" and "*clinamen*" or, as Aristotle called it, "*tuché*," the "*encounter with the real*," which "lies behind the automaton," being its very cause (Lacan, *Four Fundamental Concepts of Psychoanalysis* 53–54, 63).

57 In all my publications prior to this, I have used the term "not-all" to designate both the set of all sets (the paradox of set theory) and the female not-all (the paradox of the mathematic antinomy, as we shall see below), even as I have been aware, and arguing for, a distinction between the two structures (see *$urplus*). I would like to take this opportunity to thank the organizers and participants of the "What is Sexual Difference Now?" symposium at the University of Minnesota, in November 2009, and particularly Joan Copjec, for inspiring discussions that eventually led me to introduce a new term for the paradox of set theory in order to avoid confusion. Thus, I will be referring to the paradox of set theory with the term "all-not-all" – since, indeed, it cannot be said about the set of all sets whether it is all or not (or, what amounts to the same, it is both at once) – while reserving for the female sex the term "not-all," which is indeed defined by the characteristic, in Lacan's words, that it "will not allow for any universality – it will be a not-whole [*pas-tout*]" (*Encore* 80; Jacques Lacan, *Livre XX: Encore*, ed. Jacques-Alain Miller (Paris: Seuil, 1975) 74). In the present essay I generally take the opportunity to refine the relevant argument I made in the aforementioned book (*$urplus* 93–108); for a

different approach to the Lacanian formulas of sexuation and Copjec's account thereof, with a focus on the relation between sexual difference and genre, see my "*Genre* with and beyond Gender and Sex (a Psychoanalytic Intervention)," *Angelaki* 13.2 (2008): 93–107, but also the entire journal issue.

58 As Deleuze points out, one of the criteria of structure is that it is "a combinatory formula" of purely negative elements that "have neither extrinsic designation nor intrinsic signification," so that "the sites prevail over whatever occupies them" (Deleuze, "How do we Recognize Structuralism?" 173–74).

59 Copjec, *Imagine There's No Woman* 6.

60 Joan Copjec, "Sex is . . . ," paper presented at the "What is Sexual Difference Now?" symposium at the University of Minnesota, 13 Nov. 2009.

61 Slavoj Žižek, "The Fetish of the Party" in *Lacan, Politics, Aesthetics*, eds. Willy Apollon and Richard Feldstein (Albany: State U of New York P, 1996) 7.

62 Immanuel Kant, *Critique of Pure Reason*, ed. and trans. Paul Guyer and Allen W. Wood (Cambridge: Cambridge UP, 1998) 518; A505/B533; emphasis in the original.

63 Idem, *Prolegomena to Any Future Metaphysics that Will be Able to Come Forward as Science*, trans. James Ellington (Indianapolis: Hacket, 1977) 84–85; §53.

64 Ibid. 85; Immanuel Kant, *Prolegomena zu einer jeden künftigen Metaphysik, die als Wissenschaft wird auftreten können*, ed. Rudolf Malter (Stuttgart: Reclam, 1989) 120; §53.

65 In its extreme actualization, the dynamic or male side incapacitates the subject from simultaneously having both relations (*Beziehungen*), so that he can be either subject to the law (signifier) or free, but not both. In this case, the male rapport with the self-relation of the One can be severed, resulting in a restrictively dualistic approach via a Two that has essentially lost its connection to the One. This explains the by now standard psychoanalytic account of the male sex, according to which, as Bruce Fink puts it graphically, "a man is someone who, regardless of chromosomes, can have one or the other" – the "Other jouissance," which is "ineffable," or the "phallic jouissance," which is mediated through the signifier – "or at least thinks he can have the other by giving up the one," as in the case of "the courtly love tradition," "but not both," unlike a "woman [who] is someone who, regardless of chromosomes, can potentially have both" (Fink, "Knowledge and Jouissance" 40–41). Fink accurately notes that the "masculine structure is in certain respects synonymous in Lacan's words with obsessive neurosis" (Bruce Fink, *The Lacanian Subject: Between Language and Jouissance* (Princeton: Princeton UP, 1995) 106). But I would like to suggest that the sustenance of the rapport between the dualistic structure of the dynamic antinomy (either/or) and the structure of the paradox of set theory (all-not-all) that underlies it, *or* the foreclosure of the latter and the entrapment within absolute dualism, corresponds to the distinction between the masculine sex and obsessional neurosis, respectively. By ignoring the all-not-all in accounts of sexual difference we only reinforce the reduction of the masculine structure to obsessional neurosis, both in theory and direct experience.

66 Copjec, *Read my Desire* 8.

67 Kant, *Critique of Pure Reason* 526; A520/B548.

68 Copjec, *Read my Desire* 205.

69 The indefinite regress of the mathematic antinomy explains why, just as the masculine structure tends so easily to be conflated with obsessional neurosis (see note 65 above), the female structure is often taken as a synonym of hysteria and its endless sliding of objects of desire. The hysteric constant rejection of objects of desire ("this is not it!") and the perpetual pursuit of new ones is the extreme version of that state described by Copjec in the context of the female indefinite regress of phenomena or signifiers, in which "we are restricted to endless affirmation, that is, to affirming without end . . . the contingent series of phenomena that present themselves to us" and, hence, to negating them as necessary or absolute ("this is it!") (Copjec, *Read my Desire* 225). As with the masculine structure and obsessional neurosis, here, too, the total absorption of the subject by the not-all and the concomitant disregard (disavowal? – a term that would introduce elements of perversion) of the all-not-all of Being amounts to the reduction of woman to the hysteric, again both in theory and direct experience. In short, referring to both this and the preceding footnotes (and sexes), taken to the extreme – i.e., foreclosing (or disavowing?) Being and its all-not-all structure; for Being and relations

cannot be repressed but can be foreclosed (and, for that matter, disavowed) – sexual difference deteriorates to the difference between obsessional neurosis and hysteria.

70 Copjec, *Read my Desire* 205; emphasis mine.

71 Lacan, *Four Fundamental Concepts of Psychoanalysis* 197–98; translation slightly modified: "Ce que l'être sexué perds dans la sexualité, c'est . . . immortel. Puisque ça survit à toute division . . . la libido, en tant que pur instinct de vie c'est-à-dire de vie immortelle . . . de vie qui n'a besoin, elle, d'aucun organe, de vie . . . indestructible . . . est justement soustrait à l'être vivant de ce qu'il est soumis au cycle de la production sexuée" (Jacques Lacan, *Livre XI: Les Quatre Concepts fondamentaux de la psychanalyse*, ed. Jacques-Alain Miller (Paris: Seuil, 1973) 221).

72 For a more extensive presentation of this thesis on ethics see again my *$urplus*, where it is developed in a dialogue with the work of Alain Badiou, Žižek, and other readings of Lacanian ethics.

73 Claude Lévi-Strauss, *Elementary Structures of Kinship*, ed. Rodney Needham; trans. James Harle Bell and John Richard von Sturmer (Boston: Beacon, 1969) 24–25.

74 Kojin Karatani, *Architecture as Metaphor: Language, Number, Money*, ed. Michael Speaks; trans. Sabu Kohso (Cambridge, MA: MIT P, 1995) 95–96.

75 Ibid. 78. Castration or the prohibition of self-referentiality is that which, as Karatani writes, gives to any "self-referential formal system" – be it the One, Sex, language, or the symbolic order – an illusion of a "center that systematizes [the] system," so that it can function on the level of action. As long as this (illusion of) center is lacking "the necessity of choosing '*either* this or that' is replaced by the '*both* this and that,'" so that "the self-referential formal system is always dis-equilibrate and excessive," as it is indeed on the virtual level. Not unlike Oedipus, the prohibition of incest is "exaggerated" castration, a castration on the collective level: "Lévi-Strauss understood the elementary structure of kinship . . . as a formal structure . . . Because he had chosen the structure of kinship as his object of analysis, Lévi-Strauss had to deal with the origin of the prohibition of incest, the very structure that makes the kinship system possible. Lévi-Strauss treated the prohibition not from the genetic viewpoint," let alone the familial, "but as a logical sine qua non for the existence of a formal structure. The prohibition of incest thus became the prohibition of self-referentiality" (ibid. 93–94). For the (by Lacanian definition) "exaggerated" character of the Oedipal relation, see Deleuze and Guattari, *Kafka*, particularly 9–15.

76 Michel Foucault, *The History of Sexuality, Volume I: An Introduction*, trans. Robert Hurley (New York: Vintage, 1990), particularly the last two chapters of "Part Four" 103–31.

77 Karl Marx, *Grundrisse: Foundations of the Critique of Political Economy (Rough Draft)*, trans. Martin Nicolaus (London: Penguin and *New Left Review*, 1993) 267.

78 Paolo Virno, *A Grammar of the Multitude: For an Analysis of Contemporary Forms of Life*, trans. Isabella Bertoletti, James Cascaito, and Andrea Casson (New York: Semiotext(e), 2004) 82.

79 Ibid. 84.

80 Jacques Lacan, *Television: A Challenge to the Psychoanalytic Establishment*, ed. Joan Copjec; trans. Denis Hollier, Rosalind Krauss et al. (New York: Norton, 1990) 30.

81 Note that the present argument applies equally to images and verbal texts, once, with the advent of secular modernity, their interpretation has been set free. Free interpretation presupposes a gaze (imagined by me in the field of the Other) and, by that token, already constitutes society as specularized, at least on the virtual level.

82 Lacan, *Encore* 6; idem, *Four Fundamental Concepts of Psychoanalysis* 103–04.

83 Sartre, *Being and Nothingness* 433–34.

84 Ibid. 434; Lacan, *Encore* 10.

elizabeth grosz

THE NATURE OF SEXUAL DIFFERENCE
irigaray and darwin

I have had this strange idea for some time: perhaps the works of Luce Irigaray on the question of sexual difference – the most central concept defining her position, and one that is usually associated with her work on psychoanalytic theories of Sigmund Freud and Jacques Lacan – could find strange and unexpected support from, of all people, Charles Darwin, the father of evolution, a figure whose work also had major impact on Freud's work? Perhaps Irigaray's concept may be justly linked to that concept Darwin elaborated in his second major book, *The Descent of Man and Selection in Relation to Sex*? It may be that, ironically, Irigaray finds the greatest philosophical confirmation of her claims regarding sexual difference not so much from psychoanalytic theory but rather from Darwin's understanding of the power and force of sexual selection. And perhaps Darwin's work, with equal irony, can best be interpreted not only as a wide-ranging and systematic account of the forces that compose and transform natural existence but also as the first theoretical framework that makes the amorphous forces of sexual attraction and sexual differentiation productive of all of the richness and complexity of life?

If Irigaray sees sexual difference as the motor of cultural life, Darwin sees it as the motor of natural existence. Can Irigaray's concept, which she clearly wrote with women's social, cultural and literary subordination in mind, find resonance in biological theory? Can biology, through the transformation wrought upon it by Darwin's revolution, provide feminist thought with some useful resources by which to understand sexual difference? Is sexual difference not only one of the regulating questions of social and cultural life but also one of the questions that biology itself must address, one of the natural provocations for complexifying and proliferating life itself? Is sexual difference the universal question that life, in all its various human and non-human forms, attempts to address? Or is sexual difference one among many cultural differences, like race, class, ethnicity, or religion, that constitute the richness and conflict that characterize only human social and cultural life? What is the ontological status of sexual difference, a concept that many feminists, me included, have struggled long and hard to affirm as fundamentally cultural and variable, rather than essential, if it is rooted in and elaborated through biology? Can Irigaray and Darwin be used to illuminate the most radical insights of each other?

I want to address this conundrum: is sexual difference biological? And if it is biological, at least in part, does this mitigate its status as open-ended, as historically and geographically variable and unpredictable? Does this imply that sexual difference is not a self-contained series of pre-given differences, readily separable from all other forms of differentiation, but one embedded in all other differences? What is the status of sexual difference, that concept that Irigaray develops in her analysis of social and conceptual life, in the world of nature? And in the world of culture? And in the transition between nature and culture that is so central to most forms of political theory? In this project, I propose to use Irigaray and Darwin to confront each other's concepts of sexual difference, to open Darwinian interpretation up to feminist readings whose objectives, unlike those of Darwin's feminist contemporaries, are not the equality of the two sexes but a more adequate recognition of their differences, their modes of autonomy and even their radical incommensurabilities; as well as to open up Irigaray's work, which has focused primarily on cultural, political and linguistic differences, to new kinds of questions about nature and the animal that she has not addressed but which may enable a richer concept of sexual difference to be elaborated. I want to produce a kind of cross-mutation, an unexpected evolutionary effect of the coupling of the two, Darwin and Irigaray. Here I want to address sexual difference not as imaginary (as a relation of self and other) or symbolic (a social, linguistic and legal relation) but as real, in precisely Lacan's sense, that is, as part of the forces of the real that pre-exist both the human individual and the social order and that exceed the social order in so far as they imply problems, provocations that every form of symbolic order must address. I am interested here in the ontology, the real, of sexual difference, its force as a mode of biological and cultural dispersion rather than its imaginary and symbolic representations.

So I want to spend some time outlining Irigaray's basic precepts regarding sexual difference, explaining this concept as ontological, political and ethical principle; then to elaborate some of the key elements of a Darwinian account of sexual selection in order to show their resonances, their capacity to inflect, transform and open out our interpretations of each so that Darwin's work can be understood not only in biological but also (in a non-reductivist way) in cultural and social terms, and Irigaray's work can be understood in its full force as a dominant principle of nature and of life as well as the formative force in social and cultural life.

I irigaray and the concept of sexual difference

Irigaray's understanding of the concept of sexual difference is by now quite well known, even if, even now, some thirty years or more after its elaboration in her earlier works, it is still not very well understood. This concept is, in my opinion, the most central concept of contemporary feminist theory, the concept that elaborates both an entire research paradigm that can affect all forms of knowledge, and a politico-ethical project that involves major transformations in social, cultural and interpersonal life. It is a concept that has the potential to change how life is understood and lived, a concept that can affect how we understand both nature and culture, both ourselves and the world.

Irigaray has argued, most persuasively, that sexual difference is the threshold concept of our age, the singular philosophical issue that defines the present:

> Sexual difference is one of the major philosophical issues, if not the issue, of our age. According to Heidegger, each age has one issue to think through, and one only. Sexual difference is probably the issue in our time which could be our "salvation" if we thought it through [...] Sexual difference would constitute the horizon of worlds more fecund than any known to date – at least in the West – and without reducing fecundity to the reproduction of bodies and flesh.[1]

For her, it is not simply that sexual difference is the concept of most interest to women, to feminists, to activists involved in women's struggles: rather, her claim is stronger – that sexual difference is the most significant *philosophical* concept, the most significant thought, issue, idea, of our age, the concept that defines the social, political and intellectual preoccupations of our era. By its careful articulation, through its entwinement with all the other concepts it is bound up with and affects – all those concepts related to all lived differences, among them, differences in sexual orientation, in racial or ethnic terms, religious, economic, geographic and political difference, through differences generally inassimilable within the forms of democracy that we currently recognize – sexual difference marks the threshold of a new way of understanding ourselves, the world, and conceptuality itself. It is the pivotal concept for understanding the entwinement of all other social, political and individual differences and the bonds that may serve to unite subjects across and in recognition of these differences. This is why it is not just one concept among many others but a defining concept, a concept that opens up conceptuality itself, a *philosophical* concept par excellence, which makes it also a concept that affects life, that affects the social and all its products, that affects also what is larger than life, the social but also the natural and the divine. For Irigaray, sexual difference, as that which has been repressed by patriarchal cultures, is the concept whose elaboration has the potential to transform our relations to ourselves, to our world and to our future. Along with the concept of difference itself, which I believe is the constitutive concept of twentieth-century philosophy, sexual difference is the engine of all living difference, the concept whose elaboration has helped to specify the research paradigms and forms of conceptuality that mark the present. Whether sexual difference is an elaboration and specification of pure difference, following the contrary works of Saussure, Derrida or Deleuze, or whether pure difference is itself the consequence of sexual difference, as Irigaray implies, is a question that I cannot directly address here, but one that has marked the sometimes terse relations between Irigaray's work and that of contemporary (male) philosophy.[2]

I would like to develop Irigaray's account of sexual difference only in outline form, recapitulating many of her central claims regarding the concept, claims that have been elaborated in considerable detail in the primary texts of Irigaray, as well as through her most astute readers (Whitford; Burke; Mortensen; Schor):[3]

(1) Sexual difference is the most basic, irreducible, non-reciprocal difference between the sexes, the incapacity of one sex to step into the body, role and position of the other sex.[4]
(2) Sexual difference is morphological difference, the difference in body, in the significance and meaning of the body, and in the perceptual and qualitative immersion in the world that is developed through the body. Where many feminists have interpreted this bodily difference as anatomical and thus as given, Irigaray has insisted, instead, that bodily difference is lived, is never a raw nature but is always mediated by cultural and psychic significance. The bodies of men and women are never lived merely anatomically, but only ever in terms of their psycho-social value and meaning. Sexual difference is the concept that differentiates bodies, not in terms of their nature but in terms of their value and use.[5]
(3) Sexual difference is not only irreducible, it is also immeasurable, incalculable, a relation between terms that have no outside measure, no third term, no object to provide a metric by which to judge this relation or its constituents.[6]
(4) Thus, sexual difference is not a comparative relation between two entities, two sexes, that are independently given, like apples and oranges: it is not a comparison or contrast of two autonomous entities but is relation that is constitutive of the two sexes, which do not pre-exist their differentiation.
(5) Sexual difference does not exist in its own terms, or in terms adequate to its conceptual and political expression. Given that recorded history is the history of various types of patriarchy, sexual difference has been reduced to forms of opposition, in which man and his associated, masculinized qualities are regarded as positive and woman and her associated feminized qualities are regarded as the negation of these positive terms; to sexual complementarity, in which the feminine is only ever regarded

as that which complements the masculine rather than that which itself requires complementing; or to sexual equality, in which women and the feminine are regarded as versions of, or formally the same as, men and masculinity. In all these forms of patriarchal reduction of femininity to the needs and interests of masculinity, sexual difference is reduced to opposition, complementarity or sameness, the only forms under which the feminine is represented in patriarchal social and representational systems.

(6) Thus, sexual difference is not based on existing properties, qualities or characteristics of the two sexes, which at best reflect the social constraints of patriarchy imposed on one sex for the interests of the other, but is indeterminable, is what does not yet exist, what has nevertheless the right to exist and elaborate itself. Sexual difference has only existed in its reduced form as forms of sameness, opposition or complementarity between the two sexes, in which woman has been understood only in some relation to man as a known entity (as more or less equal, more or less dependent, more or less autonomous relative to a norm provided by men). Sexual difference is indeterminable difference, the difference between two beings who do not yet exist, who are in the process of becoming. It is a difference that is always in the process of differentiating itself.

(7) Sexual difference is both a form of difference, a mode of differentiation of that which must differentiate itself, but also a form of sexuality, a mode of erotic encounter that links different bodies in specific if open-ended modes of intensity that may result in reproduction but is not directed to it. Sexual difference, as bodily difference, is not reducible to genital differences but does include such differences and the practices they enable by necessity.[7]

(8) Sexual difference is a universal. It is that which marks all of natural as well as all of cultural life; and, moreover, it marks two modes of transition in the movement from nature to culture. It is a lived universal that is the condition for the emergence of other natural and cultural differences.[8]

(9) While sexual difference characterizes the potential as well as the actual relations between the two sexes, it thus cannot be reduced to reproduction, which is its indirect product but never the telos or goal of sexual difference. Sexual difference enables the existence of two radically different beings to create a third being, irreducible to either but the product of both. This third cannot be identified with the child, who is always one of these two. This third is the creation of something new in the relation between the two, an object, quality or relation that can mediate between the two, can confirm the relation between the two.[9]

(10) Sexual difference is not contained only within the sexual identities of male and female. Rather, its implications are far ranging and touch on the real itself. Sexual difference is not simply the existence of two irreducibly different types of subject, but at least two irreducibly different perspectives, frameworks, experiences, modes of conceptualization, forms of knowledge, techniques of existence, at least two ways of undertaking any activity. The ontology of sexual difference entails sexually different epistemologies and forms of pragmatics, that is, different relations to subjects, objects and the world itself.

(11) Sexual difference is the condition for the existence of multiple worlds, not just a single shared world. Sexual difference entails that each subject not only occupies its own morphological, perceptual and associative relation to the world but that it can indirectly access other morphological, perceptual and associative relations through their capacity to address, engage with and co-occupy a shared world, a world other than the one immediately available to the subject, through its relation to the other. The one who is sexually other than me is the one who offers me a world other than the

one I occupy, who opens up new worlds to me.

(12) Arguably, Irigaray's most contentious claim is that sexual difference is the engine or force involved in the production of all other differences, and thus has an ontological status that is radically different from that of racial, ethnic, religious, class and other differences, for sexual difference is both the universal accompaniment of all other lived differences and is one of the means for their transmission and propagation. None of these other differences has the same relation to the transition from nature to culture – all these other differences are social and cultural – and none of these other forms of social discrimination can propagate themselves without the cooperation of sexual difference.[10]

While this outline has no doubt reduced Irigaray's conception to its most elementary formulation, this exercise in outlining the various components or facets of this dazzling gem of a concept may help to explain Irigaray's hostility to those egalitarian projects that have marked much of feminist theory and practice. Any egalitarian project, whether directed to the equalization of relations between the sexes, or between races, classes or ethnicities, is, for Irigaray, antagonistic to the project of specification of differences. Egalitarian projects entail a neutral measure for the attainment of equality, a measure that invariably reflects the value of the dominant position. Egalitarianism entails becoming equal to a given term, ideal or value. Irigaray's work on sexual difference, along with the writings of other feminists and anti-racists focused on the work of specifying non-reducible differences, problematizes any given norm by which sexes or races can be measured independent of the sexes and races thus measured, problematizing the belief that there are independent variables by which equality can be charted. It is her anti-egalitarianism, her anti-essentialism and her refusal to privilege the present and the actual over the future and the virtual that marks Irigaray's unique and ongoing contribution to philosophy, and that are key elements of her understanding of sexual difference.

2 nature and culture

Sexual difference is what characterizes the natural world, the multiple forms of culture, and the varieties of transition from nature to culture. This is why, for Irigaray, sexual difference is given, not constructed. Yet even as it is given, it must also be lived, created, invented. Her position is complicated, and anticipates well in advance many objections from those who can be construed as social constructionist. We need not understand nature as static or fixed in order to understand that sexual difference characterizes nature and is one of the most striking features of the natural world. Sexual difference is a feature of all forms of culture, a problem, if you will, that each culture has no choice but to address, as it must also by necessity address the problem of mortality and the problem of cultural transmission, of how to transmit ways of living from one generation to the next. These are the biological contingencies that nevertheless become cultural necessities. Irigaray understands that nature itself provides no limit to the social and cultural possibilities of women, only the ways in which nature has been reduced to narrow biologistic explanations. For her, the problem is not biology but the ways in which biology has been dominated by masculinist thought: "what has served to exploit women is a biology interpreted in terms more masculine than feminine."[11]

If we are both natural and cultural beings, if culture is not the supersession and overcoming of nature but instead a co-existence, a mode of mutual engagement and elaboration, the cultural a mode of addressing the natural, and the natural a condition for cultural emergence, then we need other ways to understand nature than that which we abandon or move beyond. We need a new, dynamized conception of nature, an understanding of nature that acknowledges both that nature itself is continually changing, the result of change and thus never static or fixed, and a mode of production of change (and thus produces nothing fixed, nothing static or unchanging: nature is itself historical rather than anti-historical); and that nature is itself always sexed, that sexual difference marks the world of living things, plant, animal and human, or that nature itself is at least two:

> The natural is at least two: male and female. All the speculation about overcoming the natural in the universal forgets that nature is not *one*. In order to go beyond – assuming this is necessary – we should make reality the point of departure; it is *two* (a *two* containing in turn secondary differences: smaller/larger, younger/older, for instance). The universal has been thought as one, thought on the basis of *one*. But this *one* does not exist.
>
> If this *one* does not exist, limit is therefore inscribed in nature itself. Before the question of the need to surpass nature arises, it has to be made apparent that it is *two*. This two inscribes finitude in the natural itself. No one nature can claim to correspond to the whole of the natural. There is no "Nature" as a singular entity.[12]

Irigaray is seeking a new conception of nature, one very different to that found in the history of Western philosophy, that instead of seeking in nature a point of origin or departure for the social, sees in nature itself the site of productivity. If nature is never one but always at least two, and if it is a mode of becoming rather than a form of being, a mode of temporal change rather than a form of fixity or the unchanging, it may, ironically in the light of this dominant history of Western thought, provide a new mode of thinking reason, conceptuality itself, if we return to a dynamic and open-ended nature, a nature that, while universal, while providing universal questions for culture to address, provides no answers, only modes of elaboration and development. What Irigaray is interested in elaborating in a new understanding of nature is the creation of quite other models for the movement from nature to culture than those violent forces of mastery, containment and control posed by masculinist sciences, technologies and economies:

> Thus it is from the natural that we should start over in order to refound reason [...] The natural, aside from the diversity of its incarnations or ways of appearing, is at least *two*: male and female. This division is not secondary nor unique to human kind. It cuts across all realms of the living which, without it, would not exist. Without sexual difference, there would be no life on earth. It is the manifestation of the condition for the production and reproduction of life.[13]

For Irigaray, the political and cultural task of sexual difference is to become what I am, to socially and conceptually cultivate the being that is given naturally but only developed and elaborated culturally, to create a way of living that opens up and develops that nature that I am and can become. Culture is not the overcoming or rewriting of nature but its cultivation, its enhancement and expansion: culture can be, must be, more than nature's reduction to (deadly) commodity. Through a more adequate recognition of sexual difference, culture is the opening up of rather than the containment, mastery and control of nature: "My project is regulated on the basis of my natural identity. The intention is to assure its cultivation so that I may become who I am. Equally, it is to spiritualize my nature in order to create with the other."[14]

A new kind of relation between the sexes is only possible if the natural is reconceived in terms other than those which have reduced it to a frozen set of archetypes: the relations constituting the social order – interpersonal relations, relations of production and creation – are themselves founded on a misrecognized nature, a nature whose openness has been misunderstood. A new series of social relations that more adequately recognize sexual difference involve a new understanding of nature and of the foundational relation between nature and culture:

> The difficulties women have in gaining recognition for their social and political rights are rooted in this insufficiently thought out relation between biology and culture. At present, to deny all explanations of a biological kind – because biology has paradoxically been used to exploit women – is to deny the key to interpreting this exploitation.[15]

Irigaray recognizes that we need to return to a new and different concept of nature, not that concept that, through narrow scientific characterizations, reduces nature to human resource or what is useful to man, to a (temporarily) useful commodity, but a concept that recognizes our fundamental connection to the natural and our cultural duty to cultivate the natural. Nature offers, for Irigaray, not just a story of origins, the place from which the human begins, a place of pre-history, but a source of renewal and transformation of the cultural. Nature as the other of culture must itself be respected as the place and time by which culture and its human products renew and transform themselves:

> nature is a place of re-birth. Nature is a second mother, but it's also a sexed universe. Nature offers an alternative place for life and sharing in relation to the human world, the manufactured world. Rather than exploit it or forget it, I try to praise it, sing it.[16]

3 the transition from the one to the other

Yet, much as Irigaray affirms a positive conception of nature that in no way threatens or undermines the force and power of culture – unlike the vast majority of contemporary feminists, the anxiety about essentialism or naturalism regarding nature does not appear in her writings: instead, nature is valorized as a site of renewal and regeneration, as the source of culture and its transformations – it is the transition from nature to culture that interests her much more than the life of sexual difference within nature itself. In this, Irigaray remains invested, in spite of her other criticisms of this tradition, in phenomenology, and particularly in Hegel's understanding of the transition from nature to culture that has so influenced Marxism, structuralism, phenomenology, psychoanalysis, and existentialism. The psychoanalytical perspective on selfhood, for example, ultimately rehearses the Hegelian Dialectic's self-contortion in elaborating a concept of the subject as I. Indeed, modern philosophy itself seems deeply invested in understanding culture as a kind of second-order birth, a second-order nature that rewrites and transforms the first-order natural, the place where the natural bonds between mother and child are replaced by the loyalties to father, the law and the nation that constitute culture.

In Irigaray's case, this means that sexual difference, as naturally given, must be affirmed and cultivated in culture in order that it serve as the basis for a new social order and new modes of democracy. For her, it is clear that Western culture, and patriarchal capitalism in particular, whose rise Hegel chronicles, sees in nature only resources to be conquered, material to be converted into property. Men's labor is directed to the transformation of nature into commodities and in this process a natural relation of debt to materiality, to nature and to the maternal body – all unspoken conditions for the patriarchal subject – is left unrepresented and covered over by conquest:

> Because of its blindness to the significance of its patriarchal foundations, mankind no longer sees that the privilege of wealth originally concerns men alone [. . .] Wealth, understood as the accumulation of property through the exploitation of others, is already the result of the subjugation of one sex to the other. Capitalization is even the organizing force behind patriarchal power *per se*, through the mechanization of our sexually differentiated bodies and the injustice of dominating them.[17]

This hostility to nature, to materiality, expresses itself in both the desire for the domination of nature and in the control of men over women. If men "care little about living matter" (ibid.), converting matter wherever possible into property, creating nature as what must be dominated, controlled, mined, extended indefinitely, even to "the most distant stars" (5), this helps justify their refusal to recognize woman as other than, separate from and irreducible to men, to justify their reduction of women too to commodity form. It is here that Hegel is significant, for he recognizes, as others such as Marx who follow him do not, that the failure to adequately uplift and transform our relation to nature inevitably affects the forms our cultural productions take. Hegel recognizes, as others do not, that the failure to spiritualize nature, to effect an ethical relation to nature, as Antigone's story testifies, wrenches the entire social order:

> Men's society is built upon ownership of property. Life itself is treated like a commodity, productive capital, and possessed as a tool of labour, but not as the basis of an identity to be cultivated. Patriarchy cares little about spiritualizing sexually differentiated nature. This perverts its relationship to matter and its cultural organization. Hegel was particularly aware of this shortcoming of an ethics of our relationship with the natural world as it concerns the genders and their ancestries; Antigone is sacrificed because she pays her respects to the blood and gods of her mother by honouring her dead brother. Hegel wrote that this sacrifice hobbles the whole rest of becoming of the spirit. (16)

Hegel addresses the processes by which nature is sublated into spirit, and animal life, the life of plants and animals, the life that also characterizes sexual differences and sexual, including reproductive, relations between the sexes is sacrificed, surpassed and uplifted. The intense immediacy of sexual and family life must give way to the forces of the nation and the processes of sacrifice, death and mourning that enable the natural family to accede to the universal of spirit through war, through the wrenching of man from private family order into civic life. But for Hegel, this universal, attainable only when man leaves the family, is produced only by cutting man off from his sexual specificity and from his relation to the other sex, rendered him neutral in the process of becoming universal. It is, ironically, man's neutralization into citizen, subject, soldier, philosopher that effects the illusion of his disembodiment and the need for woman to be restricted to her (bodily/natural) function as nurturer/mother/carer for his body and those of his offspring. Spirit is attainable only at the cost of the sexed specificity of the body and a direct relation to the natural order: spirit is the overcoming of the natural, its transformation into the universal. Irigaray's task is to restore the sexed specificity of the body back into not only nature but into the processes of culture or spiritualization that transform natural need into social law: "Life can only be thought about, guaranteed, protected if we give consideration to *gender* as one constituent of the human race, not only in reproduction but also in culture, spirit" (132).

Instead of the neutralization of sexual and generational specificity that Hegel's account of the family as hinge between nature and culture entails, Irigaray seeks to recognize, as Hegel does, the sexed forms of the nature and to provide a model of culture that builds on rather than neutralizes this specificity: Irigaray seeks a universal that reflects the dual forms of nature itself:

> [Sexual difference] constitutes a living universal. It is a universal related to our real person, to his or her needs, abilities and desires [. . .] Thus, respecting the difference between woman and man is itself culture. It goes beyond natural immediacy [. . .] If man and woman respect each other as those two halves of the universe that they represent, then by recognizing the other they overcome their immediate instincts and drives. They are spiritual humans from the fact of recognizing that they do not represent the whole of the person and that the other cannot belong to them as their own property. In sexual difference, the negative as limit is present from the very fact of respecting natural reality as constitutive of the subject.
>
> Sexual difference is, as it were, the most powerful motor of a dialectic without masters or slaves.[18]

Irigaray seeks a continuity between the natural and the cultural, the private and the public, the family and the social order, not the split and antagonism that Hegel, and all of phenomenology that follows from him, creates between private life and civic identity, between the interests of sexuality and of social order, between the world of women/mothers and the world of men. Cultivation of the natural rather than its neutralization and universalization is the task of the social and especially of its agency, the family. Precisely as the point of transition between the natural and the cultural, the family is not the site of the abstraction and formalization of the subject-citizen, no matter how strong its task in producing citizens who take on the labor of the creation of spirit. Only if the family in its sexual and natural specificity remains connected to its origins can we have a spiritual or social and civic order that is embodied, sexed, that recognizes at least two sexes instead of a single neutral universal:

> This absence of any dialogue within the couple, this failure of sexual dialectic (on condition of rethinking the senses of the method), perverts the spirit of the individual, of the family, of the race. The concrete, which Hegel seeks in the individual, has its sexual dimension cut away. The individual is already *abstract*. This abstractness forces us to think of the family as an undifferentiated substance and not as the place of individualization, of a spiritual differentiation that can occur only if there is some polemic between the sexes.
>
> The suppression of this mini-war between living beings operates by reducing woman as woman to silence, by equating women as mothers with nature, and by obliging them to sit on their hands rather than act as citizens with an active, open and responsible role to play in building the city. The passage to the race has been perverted, falsified, in its relation to life.[19]

In short, we need both a new conception of nature, one that affirms, with Hegel, the sexual and other embodied differences that characterizes all of life in its rich variability; but, unlike Hegel, we need an understanding of the social order that affirms and remains in ongoing continuity with the natural rather than supersedes or overcomes it. Instead of nature being the resource for man's self-production, nature must be understood as that which provides us with two universals, that operates through a fundamental bifurcation, a means for the generation of new worlds, new connections, new forms of social relation. It is nature that provides not raw materials to be transformed but tendencies to be cultivated, to be acted upon with reflection and concern, forces to be recognized and provided adequate social representation:

> Without doubt, the most appropriate content for the universal is sexual difference. Indeed, this content is both real and universal. Sexual difference is an immediate natural given and it is a real and irreducible component of the universal. The whole of human kind is composed of women and men and nothing else. The problem of race is, in fact, a secondary problem – except from a geographical point of view? – which means we cannot see the wood for the trees, and the same goes for other cultural diversities – religious, economic and political ones.
>
> Sexual difference probably represents the most universal question we can address. Our era is faced with the task of dealing with this issue, because, across the whole world, there are, there are only, men and women.[20]

4 another nature

Irigaray has tended to understand nature through this Hegelian lens, as concrete materiality that requires refinement and abstraction to serve as the content for social and civic or spiritual relations. The natural is understood by her as that which relates to the inorganic and plant and animal environments of the human, as well as those elements of the human that partake of the vegetal and the animal. It is ironic that many of Irigaray's reservations about the concept of nature and her attempts to place the human within a mode of culture that radically separates the human from the animal are directly addressed through the upheavals in biological and naturalist thought provided by the publication of Charles Darwin's work. There is no evidence that Irigaray is interested in or has read Darwin's works, especially in the context of her conception of sexual difference. But it might be that Darwin's work may provide her with a richer and more resonant concept of the place of sexual difference in the universal than that provided by the phenomenological and structuralist traditions. Darwin affirms the fundamental continuity of the human with the animal and the vegetal, and the fundamental continuity between the natural and the cultural. There is no movement from nature to culture, for culture is regulated by the same broad principles as nature; and nature already contains many different forms of culture. In addition, he asserts the centrality of sexual difference – or more accurately, sexual selection, which I will argue is a form of sexual difference – in the life of species. It is only with the publications of Darwin's provocative picture of the web of life adapted through natural and sexual selection that the human's place in nature has finally achieved a kind of scientific and philosophical recognition that affirms man's place in nature as much as in culture. Darwin's understanding of nature supersedes Hegel's, and indeed entirely reorients the trajectory of German Romantic nature philosophy.[21] The human, the two forms of human, are given in nature, but cultivated through culture. Culture is not the completion of nature but rather the experimentation with nature's open possibilities, its possibilities for both oppression and liberation.

My project does not aim to replace Hegel's conception of nature with Darwin's – though such a project is an interesting one – but to claim that Irigaray's understanding of sexual difference as a striking characteristic of both nature and culture may be elaborated and developed through Darwin's work on species. It is my claim that a non-reductionist understanding of sexual difference as biological force finds its greatest support in Darwin's writings, and Darwin provides biology with its most open philosophical framework. To the extent that Irigaray's conception of sexual difference may be sharpened and further elaborated through Darwin's account of sexual selection, I will also claim that Darwin's work is further expanded and made relevant to feminist concerns through itself being interpreted not through feminist egalitarianism, which Darwin himself sometimes addresses in both respectful and sarcastic terms, but through a feminism of sexual difference. That is, instead of a philosophical framework in which the formal or abstract identities of male and female subjects are treated as if they are or could be the same, a recognition of the irreducible differences between the sexes – and the consequent necessary failure of any but the most abstract forms of equality – would also help to clarify the relevance of Darwin's work for feminist thought. Darwin provides perhaps the most systematic and elaborate explanation for the genesis and near-ubiquity of sexual difference – in its various forms, for it does not always take on the forms that mark the human – and thus his work finds an unexpected alliance with Irigaray's understanding of sexual difference, which may help to explain its unrecognized radicality. In other words, although the positions of Irigaray and Darwin were independently formulated and function perfectly well without the input or framework of the other discourse, nevertheless, the position of each is highlighted and intensified through the contributions provided by the other. Indeed, each is radicalized, taken further, through their association or affiliation.

Though Irigaray never refers to Darwin's work or to a conception of nature that is dynamized and fully compatible with and inseparable from culture (at least in the case of social animals) she nonetheless approximates many of his concepts. For example, she distinguishes between something like natural and sexual selection, natural necessities for life, and sexual requirements:

> Two natural necessities dominate societies. One of them may appear to be neuter, unmarked by the sexual: we all have to breathe, feed, clothe and house ourselves. Our societies are controlled by this need, which, rightful as it is, accords money a power that is totally disproportionate [. . .] In addition to need, there is another dimension in the person, that of desire, which is linked to energy, particularly sexual energy. This dimension of the person as sexed is important for social production and reproduction: without it, there is no society. Yet the dignity and necessity of sexual difference goes unrecognized.[22]

Although she does not use his language, this distinction between two natural necessities is the distinction between natural selection and sexual selection, a distinction between the struggle for existence against natural elements and chance itself; and the struggle to attract sexual partners that characterizes sexual selection. These two forces, Irigaray recognizes, are what nature bequeaths to all forms of social organization: the necessity to provide for conditions which sustain life; and the necessity of addressing sexual energy and attraction. Cultures may vary immensely in how they address these two necessities, but each must find some way in which these are adequately addressed. Patriarchy is one such attempt, but by no means the only one possible. Perhaps Darwin can provide another concept of nature than the Hegelian framework through which Irigaray understands the transition from nature to culture. And perhaps Irigaray can provide a framework for understanding the differences between male and female, neither as oppositions nor as forms of sameness, but as irreducible terms governed by relations of difference that will help clarify Darwin's understanding of sexual selection and open it up to new feminist understandings of evolution and biology.

5 darwin and feminism

The relations between feminism and Darwinism have always been complicated, ambivalent and multi-stranded. From the very outset, feminists have resisted many of the methods and questions Darwinian thought has developed and explored. Within four years of the publication of Darwin's *The Descent of Man and Selection in Relation to Sex* (1871), Antoinette Brown Blackwell published *The Sexes through Nature* (1875); Eliza Gamble published *The Evolution of Woman: An Inquiry into the Dogma of Her Inferiority to Man* (1893). Both were concerned not with Darwin's understanding of the relations between the production and inheritance of variation and the operations of natural selection, that is, with his broad understanding of evolution, with which they concurred, but rather with Darwin's understanding of sexual selection, with what they perceived to be Darwin's privileging of the male position and its concomitant values. This suspicion has marked the development of Darwinian thought for 150 years.

Feminists have had reason to be extremely wary of the eruption of new Darwinian projects that seemed to emerge at precisely the same time as the resurgence of feminist thought in the late 1960s and 1970s (E.O. Wilson's *Sociobiology: The New Synthesis* was published in 1975), and have quite justifiably wanted to distance themselves as much as possible from the essentialist and reductionist assumptions that emerged and proliferated with the eruption of sociobiology and the merging of natural with sexual selection that was effected through it. Sociobiology and its contemporary heir, evolutionary psychology, have, through their reduction of maleness and femaleness to their reproductive capacities and activities alone, become the objects of considerable debate among feminist theorists working in both the sciences and the humanities: while it is clear that there is immense resistance to sociobiological thought by many feminist scientists, it is also clear that there is a growing body of feminists, many working within evolutionary biology (and evolutionary psychology) who believe that it can provide some answers to feminists questions about the tenacity of patriarchal power relations and the effectivity of social policies addressing evolutionary effects.[23]

Sociobiologists have performed a number of philosophical reductions – the reduction of the living body and its associated questions of embryology and development to its selfish genes (for the body is but the temporary vehicle for the immortal gene),[24] the reduction of sexual selection to natural selection, the reduction of maleness and femaleness to the size of gametes,[25] the reduction of the continuity of life to algorithmic steps.[26] It is only something like a theoretical or philosophical rereading of Darwin's own texts, which remain as powerful today as they were at the time of their publication, that can restore the continuity and the picture of life as a whole, in all its elements, to our understanding of the life of species and to man as a participant in the same evolutionary schema. It is in the spirit of philosophy's capacity to think through the implications of scientific theses in ways that perhaps scientists cannot that I want to develop an analysis of the relations between natural and sexual selection in Darwin's own writings, for it is there that we find resistances to the contemporary impulse to reductionism.

I believe that if Darwin's works are carefully addressed through the concept of sexual difference that Irigaray has developed, a new and quite different understanding of sexual selection than that which dominates sociobiology emerges; Darwin's own writings already outstrip his later interpreters. The meticulous attention not only to the accumulation of vast and detailed examples as well as his focus on a large philosophical picture of the lived world is missing in the writings of many, perhaps most, of his followers. If sociobiology is one side of a carefully triangulated series of positions regarding how to understand life and its bifurcations, then perhaps feminist (and its alliances with anti-racist and post-colonial) struggles represent another position that must be addressed and complicated here, as they too were issues of direct relevance in Darwin's own texts. His writings have given us the resources to develop a counterforce to the prevailing scientism that dominates much of evolutionary thought. I am not developing a critique of science or scientific method at all – Darwin's work itself is, after all, precisely the birth of a new kind of science, a new scientific attunement to the question of history and the movement of time, that is, a major reconfiguration of how the science of life must work – but trying to ascertain what the limits of scientific relevance may be, the points beyond which some more general or theoretical reflections are necessary.

A philosophical exploration of the concept of sexual selection in Darwin's writings, directed by a commitment to a feminism of irreducible difference, would enable a new non-reductionist understanding of sexual selection as a principle vital and irreducible to natural selection. Feminism, to the extent that it is implicated in egalitarianism, tends to resist Darwin's understanding of sexual selection which has privileged maleness and attributed activity to it while affirming the relatively passive position of femaleness; yet even though Darwin provides us with a theory of the centrality of sexual dimorphism (or polymorphism!) to the production of greater and greater variation, his work has been understood in largely essentialist, biologistic or invariable terms by the vast majority of feminists, who either affirm this essentialism,[27] or in some way want to challenge it.[28] But those feminists who are committed to the concept of the irreducible difference between the sexes may find in Darwin's writings surprising confirmation of their claims, as well as a deeper understanding of the nature/culture relation.

Darwin's own relation to feminism and to women's struggles is quite complex. He is clearly sympathetic with and committed to programs of social equalization, and especially education, to breach the social gap between men and women. Along with his friend and correspondent John Stuart Mill, he believed that the index of a culture's openness is the way in which men treat women in culture. Slavery and the denigration of women are, for him, twin evils that exist in both barbarous, but also in developed, cultures:

> The great sin of Slavery has been almost universal, and slaves have often been treated in an infamous manner. As barbarians do not regard the opinion of their women, wives are commonly treated like slaves. Most savages are utterly indifferent to the sufferings of strangers, or even delight in witnessing them.[29]

He recognizes that it is patriarchal power that has commonly functioned to keep women in the state or servitude that she shares with man's animals.[30] In part, the progress of civilization can be mapped through the ways in which man treats woman: "Man is more powerful in body and mind than woman, and in the savage state he keeps her in a far more abject state of bondage than does the male of any other animal; therefore it is not surprising that he should have gained the power of selection."[31] He also believes that although man is more educated than woman, with the benefit of an intense and rigorous training in both mind and body, woman can, in addition to the task of procreation and nurturance, become as educated, as civilized and developed as man.[32] This liberal egalitarianism, his clear hostility to the rampant racism of his time and his commitment to questions of class – and indeed his relations to Marx and Engels – were pervasive political commitments of his family. It is hardly surprising that Marxists greeted his account of evolution through natural selection with the same warmth and affinity as contemporary feminists. After the publication of *The Origin of Species* in 1859, Marx wrote to Engels: "Although it is developed in the crude English style, this is the book which contains the basis in natural history for our view."[33] Indeed, Marx clearly felt a very great affinity with Darwin's work on the struggle for existence. He inscribed into Darwin's copy of Volume 1 of *Capital*, "Mr. Charles Darwin on the part of his sincere admirer Karl Marx London 16 June 1873," and offered to dedicate Volume 2 to Darwin.

Darwin's own politics was directed to an impulse to consider the sexes, all races and all classes as fundamentally equal, as more governed by degrees than by any insurmountable gap. But this may be because the concept of a politics of difference – among the most inventive creations of the twentieth century – had yet to be invented. Egalitarianism represented the highest aspirations of a culture concerned for the well-being of all its members. Yet the concept of fundamental difference, a difference in bodies, and thus in interests, in perspectives and values was already emerging in his understanding of animal existence that would provide the conceptual and historical preconditions for Irigaray's understanding of sexual difference and the proliferation of politics of racial, class and other differences and would itself benefit from being interpreted and understood from such a perspective.

6 sexual selection

Darwin carefully distinguishes sexual from natural selection. While it is natural selection that has privileged sexual reproduction over the forms of asexual reproduction that still characterize some forms of life (bacterial, viral, protozoan), sexual selection is a principle irreducible to natural selection. Natural selection is always ultimately directed by the struggle for existence, the struggle to survive: it is oriented to the factors regulating life and death. Sexual selection, by contrast, is directed to the struggle to attain sexual partners and thus the stakes are much less severe and dire:

> This depends, not on a struggle for existence, but on a struggle between the males for possession of the females; the result is not death to the unsuccessful competitor, but few or no offspring. Sexual selection is, therefore, less rigorous than natural selection.[34]

Sexual selection is aimed at attaining desirable sexual partners. While it is in some sense subordinated to the forces of natural selection, which remains the final arbiter in the assessment of the value of any characteristics, sexual selection is a principle separate from and not reducible to natural selection. This is perhaps the most major distinction between Darwin's own writings and those of contemporary sociobiology, for whom sexual selection is in fact a subtle form of natural selection, and attractiveness is an index for fitness. But for Darwin, sexual selection is not only a separate principle from natural selection, for it could have never been deduced from a knowledge of natural selection; in addition sexual selection attenuates and problematizes the criteria by which natural selection operates and substitutes its own, sometimes contrary, principles of taste, attractiveness or desire. Sexual selection not only complicates natural selection; it has the potential to imperil life, to render various activities or qualities more noticeable, more obvious and as vulnerable to attracting predators as much as potential sexual partners.

Sexual selection heightens and intensifies the manifest differences between the sexes. Paradoxically, sexual selection regulates many of the perceptible differences between the sexes but it does not direct itself to those sexual differences that lead directly to reproduction! The reproductive capacities of living bodies are regulated by natural selection and are part of the struggle for existence to the extent that they determine questions of life (and death). Thus sexual selection can refer only to those characteristics and activities that surround the reproduction but fall short of fertilization. All those organs and activities that are directed to reproduction and to the care and nourishment of the young are considered to be the products of natural selection. Thus, both the existence of differently sized gametes – sperm or ova – and of some so-called secondary sexual characteristics, such the existence and operation of breasts or abdominal sacks, are the result of natural selection: their presence has proved advantageous in sustaining life. All other sexual characteristics, those not directly related to the production and care of the young but rather to attracting sexual partners, are forms of sexual selection, even if these may also serve in some way as forms of advantage in the struggle for existence:

> The female often differs from the male in having organs for the nourishment or protection of her young, as the mammary glands of mammals, and the abdominal sacks of the marsupials. The male, also, in some few cases differs from the female in possessing analogous organs, as the receptacles for the ova possessed by the males of certain fishes, and those temporarily developed in certain male frogs. Female bees have a special apparatus for collecting and carrying pollen, and their ovipositor is modified into a sting for the defence of their larvæ and the community. In the females of many insects the ovipositor is modified in the most complex manner for the safe placing of the eggs [...] There are, however, other sexual differences quite disconnected with the primary organs with which we are more especially concerned – such as the greater size, strength, and pugnacity of the male, his weapons of offence or means of defence against rivals, his gaudy colouring and various ornaments, his power of song, and other such characters.[35]

Sexual selection explains the manifest differences between the sexes in those species where they differ in appearance (other than in sexual or reproductive organs). To the extent that the two sexes of a species face the same exigencies of life and death, when they are perceptually differentiated (either visually, sonorously, through smell or in any other discernible way), especially in secondary sexual characteristics or in other manifest differences, this is the result of sexual rather than natural selection:

> Thus it is, as I believe, that when the males and females of any animal have the same general habits of life, but differ in structure, colour, or ornament, such differences have been mainly caused by sexual selection; that is, individual males have had, in successive generations, some slight advantage over other males, in their weapons, means of defence, or charms; and have transmitted these advantages to their male offspring.[36]

Natural selection privileges some individuals and species over others in the struggle for existence; sexual selection privileges some members of one sex within a species over others in the struggle to attain desirable sexual partners. Sexual selection tends to differentiate the sexes more and more from each other in appearance. That these are forms of sexual selection rather than natural selection is clear from the fact that both male and females survive equally well though they look and act in increasingly divergent ways over the passage of time. And the less attractive members of either sex would continue to reproduce in lieu of the presence of more attractive members. Sexual selection produces characteristics and activities that are linked to both appeal and attraction and to spectacle and display.[37]

Sexual selection generally takes two forms, which are not mutually exclusive. On the one hand, it consists in various battles or forms of competition between members of the same sex, usually males (or more rarely, females) for the right to select the sexual partner who appeals most to them. These various forms of competition between males – whether these are contests of strength and prowess, singing competitions, the creation of the most appealing nest, the arriving well before one's rivals at a mating location, or forms of strutting and display, produce a productive spiral of intensification of appealing characteristics (for not only are these characteristics heritable, so too is the finding of these characteristics attractive). The most successful males competing with other males are not always the most successful males in attracting female interest, but commonly they are.[38] Male competition intensifies and stimulates the creation and increasing elaboration of erotically desirable characteristics, whether these are linked to strength and agility, pugnacity and vigor, singing abilities, beauty of plumage or scales, upright stance, or the elegance of their built structures. In competing with each other for primacy, there is a struggle among various forms of strength and attractiveness that intensifies and proliferates the successful bodily forms and forces, actively transforming the appearance and capacities of such competing individuals (and their offspring). This may even result in more success in the struggle for existence: "the advantages which favoured males have derived from conquering other males in battle or courtship, and thus leaving a numerous progeny, have been in the long run greater than those derived from rather more perfect adaptation to the external conditions of life" (279).

On the other hand, sexual selection also entails forms of usually female discernment, in which females select from a number of possible partners those which most appeal to them. This distinction between as it were active male competition and passive forms of female discernment has, not surprisingly, been the object of much (egalitarian) feminist criticism, for it seems to reproduce precisely the most stereotyped images of male and female as oppositional, active or positive and passive or negative, terms. In fact, however, Darwin devotes considerable detail to the analysis of female competition and male discernment, which seems more common in insects, fish and some species of birds than in higher mammals (see, for example, 272). For him, this is an empirical question, a question determined by the observation of the activities of various species rather than an a priori assumption. In any case, for those patriarchs contemporary with Darwin, the very idea of female discernment was disturbing, for it assumes a degree of intelligence, of preference and choice that was at odds with the assumption that it is males who are the primary force of sexual encounters and female preferences are of little consequence (such too are the current forms of sociobiological justification of rape as evolutionary tool: it assumes that the only sexual forces are active forces and that these are male). Darwin makes a strong argument in favor of female selection: only female discernment can explain the increasingly different bodily forms of males relative to females, only the advantages on males imposed by their attractiveness to females can intensify and exaggerate these qualities in successive generations. Only female discernment or taste can explain the ongoing existence of extravagant, sometimes even endangering, ornament:

> Does the male parade his charms with so much pomp and rivalry for no purpose? Are we not justified in believing that the female exerts a choice, and that she receives the addresses of the male who pleases her most? It is not probable that she consciously deliberates; but she is most excited or attracted by the most beautiful, or melodious, or gallant males. (123)

Females are, generally, both less eager and more discerning in their sexual encounters than males of the same species. They function to consider sexual encounters rather than to immediately enact them:

> The female [. . .] with the rarest exception, is less eager than the male. As the illustrious Hunter long ago observed, she generally "requires to be courted"; she is coy, and may often be seen endeavouring for a long time to escape from the male. Every one who has attended to the habits of animals will be able to call to mind instances of this kind. Judging from various facts, hereafter to be given, and from the results which may fairly be attributed to sexual selection, the female, though comparatively passive, generally exerts some choice and accepts one male in preference to others. Or she may accept, as appearances would sometimes lead us to believe, not the male which is the most attractive to her, but the one which is the least distasteful. The exertion of some choice on the part of the female seems almost as general a law as the eagerness of the male. (273)

It is significant that Darwin argues that if male competition intensifies the physical capacities, strength, energy, agility and war-like activities of males, it is female discernment that intensifies male appearance, the production and extension of ornaments, forms of charm and beauty. Female discernment intensifies male beauty and attractiveness, sometimes even at the cost of male survival:

> It is certain that the females occasionally exhibit, from unknown causes, the strongest antipathies and preferences for particular males. When the sexes differ in colour or in other ornaments, the males with rare exceptions are the most highly decorated, either permanently or temporarily during the breeding-season. They sedulously display their

> various ornaments, exert their voices, and perform strange antics in the presence of the females. Even well-armed males, who, it might have been thought, would have altogether depended for success on the law of battle, are in most cases highly ornamented; and their ornaments have been acquired at the expense of some loss of power. In other cases ornaments have been acquired, at the cost of increased risk from birds and beasts of prey. (123)

Darwin evinces a number of arguments for the distinctness of sexual selection from natural selection. The attributes produced as a result of sexual selection

(1) are much more marked in adults than in the young; particularly the qualities of adult maleness are often attenuated or difficult to observe in the young and only emerge after sexual maturity (this is what enables Darwin to suggest that human male hairiness, and especially the beard, are consequences of sexual selection, and thus exert an appeal for human females; 316);
(2) tend to be inherited by offspring of the same sex and are thus heritable: males tend to inherit those characteristics that mark their successful male progenitors; female, the same with their female progenitors. Although Darwin was unaware of genes and their role in the inheritance of characteristics, he was acutely aware of the phenotypic forms of such characteristics. Without their tendency to be inherited, sexual selection would not accumulate characteristics but would function only for the current generation;
(3) become increasingly intensified as time progresses, that is, the sexes are less and less alike with the passage of time (this forms the basis of Darwin's assumption that sexual dimorphism emerges as a powerful evolutionary strategy to supersede or surpass hermaphroditic or single-sexed reproduction);
(4) not only characterize many perceptible qualities of the body but also character traits, forms of personality and modes of activity, that is, it functions with both the products of body and those of mind; and
(5) function primarily through subjective qualities, taste, appeal, what is attractive, alluring for its own sake. This may coincide with a discernment of fitness but it may not. Sexual selection elevates the artistic, the gratuitous, the ornamental, for its own sake, for the sake of pleasure or beauty, even to the point of imperiling the more beautiful and noticeable individuals over their less beautiful competitors. Sexual selection both augments and problematizes natural selection.

It is because the concept of sexual selection in Darwin's writings is so closely linked to taste, the discernment of beauty and the appeal of the ornamental, the secondary, the frivolous, that there has been an enormous investment in contemporary evolutionary theory to explain sexual selection as a form of unconscious or indirect natural selection. But Darwin is right to separate the (relative) usefulness of those attributes acquired through natural selection for they function according to contrary principles and have very different effects. Sexual selection is primarily creative: it enhances, elaborates and exaggerates individual differences, to make them as enticing and appealing as possible. It complicates and intensifies attractive individual differences; natural selection, by contrast is primarily negative: it tests and eliminates the less fit rather than privileges the more fit. Sexual selection relies on altogether other criteria than those regulating natural selection, not the impersonal criteria of random chance, but the personal criteria of attractiveness. Darwin suggests that if natural selection directs those qualities, those differences that privilege survival, it is sexual selection that maximizes difference, that generates individual variation and that guarantees that offspring will be different from their parents, even as they share certain inherited qualities from them. Sexual selection is the engine for the creation of those differences that natural selection evaluates: it is the generator of individual variations that are then assessed by natural selection.

7 the origin of sexual selection

While the origins and emergence of sexual selection occupied only a few pages of the *Descent* it is a question that clearly fascinated Darwin.[39] In spite of his reserving a detailed discussion of sexual selection until the publication of *Descent* in 1871, sexual selection was in fact one of Darwin's earliest hypotheses, the one that serves to distinguish him most clearly from the many other naturalists also working on the mutability of the species at roughly the same time. Already in his travels on the *Beagle* he was fascinated with those organs, appendages and activities that seem to have no direct bearing on an individual's capacity to survive but seem to be linked to sexual attraction. In his third and fourth Notebooks he addresses the question of the origin of sexual dimorphism. And in his joint publication with Wallace, *Evolution by Natural Selection*, where he first publicly presents his research on natural selection,[40] while Wallace makes no mention of sexual selection, Darwin "devotes a disproportionately large section to it,"[41] which implies that he both wanted to emphasize its importance as well as to mark out in advance his differences from Wallace's position.

As is well known, Darwin held off publication of both *Origin* and *Descent* for a number of years, in fear of the likely reception of these works and in the belief that he needed to demonstrate his empirical and observational credentials before presenting his quite wild and clearly philosophically oriented understanding of the movements that constitute evolution. What is perhaps less well known is that, after he had drafted an early version of *Origin* but before he published it, and well before the publication of *Descent*, he devoted a mystifyingly immense amount of time – eight years – to the thorough analysis of a lowly creature, the barnacle. Rather than seeing this as a distraction from the anxieties surrounding the publication of what he understood would be the profoundly contentious claims of the transmutation of species, his work on barnacles, which admittedly took a good deal longer than Darwin had expected, was in part directed to major revisions in the taxonomy and categorization of barnacles, but largely directed to an analysis of the origins of sexual selection and the very peculiar forms of sexuality that emerged from his laborious observation, dissection and analysis of the vast range of living and fossil barnacles. In this process, he became the world authority on barnacles and his four monographs on the topic remain even today an "indispensable reference" for those interested in the topic.[42]

Darwin does discuss the possible origins of sexual bifurcation briefly in *Descent* where he speculates on the origin of certain vestigial or rudimentary organs of the one sex being found in the other sex. This is quite striking in the case of the embryos of many species, but there are also clearly traces of the other sex, such as nipples in males, that occur in the mature individuals.[43] The existence of such organs attests to "some extremely remote progenitor of the whole vertebrate kingdom [that] appears to have been hermaphrodite or androgynous."[44] The earliest ancestors of sexually bifurcated species, he speculates, may have included the reproductive organs of both sexes within a single form. Mammalian life may have descended from earlier forms that were organically bisexual. The sexes divided well before mammals as a category were distinguished from their progenitors[45] and each sex still carries the rudiments of its preceding pre-mammalian androgynous state. It is perhaps his fascination with this question of the origins of sexual difference, raised so briefly in *Descent*, that directed him to the laborious investigation, by no means a detour, of the evolutionary history of the barnacle.

Barnacles have many peculiarities – their anatomy and morphology, certainly their variety and scope, and their amazing success in surviving in a wide range of environment, virtually on every shoreline – which were largely unknown when Darwin began his researches. One question that fascinated him was the genealogical relation between barnacles and other crustaceans. Barnacles were insensibly differentiated from crustaceans over vast periods of time, and Darwin hypothesized that their capacity to secrete a glue-like substance which enables them to cement themselves onto various surfaces, one of their characteristic but not universal features in the present, seems to derive from the presence of a sticky substance, present in crustaceans, that lines the female tract. Darwin suggested that the contemporary barnacle was descended from an ancestral barnacle that was hermaphroditic, which may explain how barnacles had acquired their cementing capacities, but would not explain how male barnacles could attach themselves either to objects, as do the females, or, most importantly, to females in order to facilitate reproduction. Much of Darwin's detailed work was based on providing intermediate examples, drawn from fossilized barnacles, that could explain the sequences of anatomical transmutation that transform a hermaphroditic ancestor into two sexes. What Darwin discovered was that many of the barnacle fossils, especially the oldest, exhibit an hermaphroditic anatomy, but, more peculiarly and surprisingly, this is also true of some quite rare contemporary forms of barnacle, which appear to be hermaphrodites, though the male organs are in a process of atrophy and are "microscopically small." In addition, these largely female hermaphrodites have various small parasites attached to them that are, after microscopic inspection, dwarf males, males in the processes of emerging as a separate sex but which are still largely primitive modes of insemination, little more than sacks of sperm.[46] In one and the same species, then, there are hermaphrodites, which are largely female, there are atrophying male organs, and tiny self-contained males. Although hermaphrodites (both in the plant kingdom as well as among various sea creatures) have the possibility of self-fertilization, Darwin has argued that this rarely occurs; hermaphrodites generally cross-fertilize. They require the close proximity of other hermaphrodites/females and males.

This arrangement of hermaphrodites and complemental males in the case of barnacles seems to make sense of the morphological limits of barnacle life. Sessile or immovable animals, such as the barnacle (or Darwin's other favorite, corals) grow their shells directly onto a fixed place, and after this attachment they are not able to move. Thus barnacles can only exchange sperm with their nearest neighbors. If their nearest neighbor is one of two sexes, then there is only a 50 percent chance that a male and female will be in proximity with each other. With two hermaphroditic individuals, the possibilities for cross-fertilization are greatly enhanced for any two individuals in proximity. And the emergence of complemental or parasitic males, males which attach themselves to the female/hermaphrodite to live parasitically, is also much more likely to result in cross-fertilization than two separate and autonomous sexes. In addition, male barnacles, even those of dwarf stature, are known to have extraordinarily long penises (the largest penis size to body ratio of any animal[47]), to maximize the chances of fertilization. Having both types of morphologies only further guarantees the likelihood of reproduction. This indeed may explain the remarkable evolutionary stability of barnacles, which, while they have well over 1,200 different species – species which differ widely in their anatomical structure and in their reproductive relations – all still bear a striking resemblance to their most primitive ancestors.

As primitive sea creatures, barnacles are very much like the earliest forms of life to appear after the emergence of animals from plants (and their bisexual structure attests to this). Yet barnacles are also pervasive, along every coastline and tidal location across the globe. Some barnacle fossils are as old as 500 million years, while many have stable forms that can be dated back 20 million years. Darwin hypothesizes a genealogy of contemporary barnacle forms from an originally hermaphroditic barnacle ancestor, an ancestor that emerged by slow degrees of change from the crustaceans as distinctively and uniquely hermaphroditic. The question was: how does this hermaphroditic creature become a sexually differentiated creature?

In his exploration of the stalked barnacle *Ibla*, common in the Philippines, Darwin discovered only females in all of his dissections but no males. It was only when he turned his attention to the tiny parasites on the body of the female *Ibla* that he understood that these parasites were tiny males, very primitive creatures with no mouth, no digestive system, little more than living tubes of sperm, attached to or burrowed into the female's body. Yet when he dissected Australian *Ibla*, he discovered, along with females, some hermaphroditic specimens. Both females and hermaphrodites had complemental males burrowed into their bodies, which in no way resembled either the females or the hermaphrodite morphologies. These resembled no other animal forms, though they did resemble species "in the Vegetable Kingdom."[48] The *Ibla* provided a concrete illustration of the sequence of evolutionary elaboration. First are the ancient progenitors of today's vast range of barnacles and barnacle forms, hermaphroditic barnacles, which are still quite prevalent today among the various species of barnacles. Then emerged barnacles like the *Ibla* which represent a transitional stage: here hermaphrodites incorporate male organs that are clearly in the process of atrophy, as well as robust female organs, for their function can now be assured fertilization with the emergence of the complemental males. Eventually these hermaphroditic male organs will either disappear or become vestigial (much like the atrophied stamen and pistils of hermaphroditic plants). Through gradually imperceptible changes come separately sexed barnacles, the females resulting from the hermaphroditic forms, and the males emerging as complemental. These complemental males are perhaps not truly autonomous, for they usually have no modes of sustenance, ingestion or digestion. They are neither entirely autonomous nor entirely submerged in and part of the female/hermaphroditic body.[49] These species are currently in the slow process of transforming from hermaphroditic to bisexual and then to two separate sexes which emerge with stalked barnacles.[50]

What emerges from his analysis of barnacles is the story of the half-emergence of maleness, not femaleness, whose reproductive capacity must be marked somewhere in the living being. The significant question is less how living beings (plants and animals) reproduce themselves, for in order for life to emerge at all, the (female) capacity for generation must be assumed. The question is really: why is there a second force of generation? Why is there maleness? What advantages does the emergence of a separate form of maleness create? Why, in other words, does sexual selection erupt? What advantage does sexual selection bring such that it generates more than one body type, more than one form of inheritance?

For Darwin, the answer is clear. Sexual bifurcation, the eruption of more than one (by default, female) sex is a strategy to maximize the potential for variation, to maximize the forms of living beings, to maximize difference itself. Natural selection selected the strategy of dividing the sexes into (at least) two bodies rather than a single morphological type – not for all forms of life, especially those invested in evolutionarily stable environments, but for those involved in changing situations. Sexual selection is the most reliable form of reproductive strategy in a large number of contexts, for plants and for animals, because it provides the conditions under which the greatest variety of living beings can be produced from which the fittest, or the most contextually embedded, can be selected to produce the next generation in greater numbers than their less fit or embedded counterparts. Sexual bifurcation is privileged by natural selection as a means for maximizing survival of some, if not all, of the members of a particular species, that is, as a mode of differentiation that guarantees the maximization of differences between individuals. And in turn, sexual selection then functions to deflect natural selection through its extravagant and excessive pleasures, its inventions and intensifications of new relations, new forms of attraction and new modes of artfulness. Sexual selection is arguably the greatest invention of natural selection.

Darwin claims that it is responsible not only for the vast range of variations of life on earth but more, it is responsible for the creation of a kind of arms war that intensifies the value of certain qualities, those that are sexually alluring and attractive, and gives them a disproportionate value over other qualities that may not be warranted in terms of natural selection alone. It is responsible for the intensification of beauty over generations and for the proliferation of colors, sounds, forms that are pleasing to members of one's own species if not to others. Sexual selection has, Darwin even suggests, played a powerful role in the creation of the human, which has its present form not only as a result of biological survival strategies but even more through the forms of appeal and pleasure that account for many of the perceptible features we still find appealing today, those linked to racial differences, to degrees of hairiness, to the timbre of the voice, to height, strength, grace and other tangible qualities. Indeed, Darwin even suggests that it is sexual rather than natural selection that is responsible for the two most highly regarded attributes, the human capacity for speech (which is for Darwin derived from the human attraction to musical cadences) and the remarkable growth of human cerebral capacities, or intelligence.[51]

8 sexual difference as sexual selection

So is Darwin in fact the first theorist, the first feminist, of sexual difference? Although clearly for him sexual difference is not irreducible insofar as it is derived in a slow, even imperceptible movement from anterior forms of existence, nevertheless sexual difference, di- or polymorphism, once it erupts as a random invention of life, comes to characterize most of life in increasingly marked terms. Once hermaphroditic or female forms elaborate the possibility of other morphologies of the same species as their forms of variation, sexual difference is increasingly marked, emphasized and each sex is sent on its own specific trajectory, never to be reconciled in a single entity again. Sexual difference is the random development that alters the course of life as we know it, deflecting all other forms of evaluation and selection through the inexplicable, unpredictable, incalculable vagaries of taste, desire, appeal.

Sexual bifurcation establishes a challenge, a problem for all forms of life: how to engender life, given that life is no longer self-perpetuating, that life, whether natural or social, now requires at least two. How to engender sexual attraction, sexual selection and the production, with variation, of new generations. How to entice the creation of new generations. How to live together, at those times and in those forms that one does, given the differences in morphology, and with them, in organs, activities and practices. What kinds of relations of co-existence, whether collective or individual, facilitate such engagements? Darwin addressed these questions very carefully, slowly and with immense detail because it is not only in the world of animal existence that they are relevant: these are, of course, also the most basic questions for human life. Darwin's reluctance to address human sexual selection in anything but the briefest terms, addressing only the broadest descriptive differences between the sexes, is by now well known. This was much less his terrain of expertise than the activities and physiologies of animals, even though some of his other work – his notes on the early development of his own children, his detailed observation of his daughter Annie's prolonged illness, as well as his writings on the expression of emotions – evince his astute powers of observation and reflection in the field of human affairs as well. Yet he largely refrains from much speculation on human sexual relations and human forms of attraction (though these do, for him, provide a detailed explanation of racial differences).

Irigaray, of course, has no such hesitation: the domain of human relations, and especially the relations – sexual, domestic, economic, linguistic – between the sexes are the objects of her intense analysis throughout her writings. Although she does occasionally mention animal and plant relations, she addresses them only in passing and only in relation to how they illuminate or alleviate the position of woman in phallocentric cultures.[52] There can be no direct reconciliation between Irigaray and Darwin to the extent that each is implicated in quite different projects, Irigaray in philosophy, and addressing only the human, Darwin in biology, and addressing primarily the animal. Yet it may be that by juxtaposing or cross-fertilizing their quite different contributions, the work of each can be sharpened, made more conceptually incisive, more broadly relevant.

If Darwin's work can be understood not in the context of egalitarianism, which defined the horizon of radical thought in the sexually and class-stratified society of his time, but through Irigaray's understanding of sexual difference, then it will be broadened and made relevant for contemporary political accounts of difference. Instead of there being a neutral norm by which both sexes can be socially assessed, Darwin's work can be understood, well beyond the racist legacy of eugenics or artificial selection, as a profound mode of analysis of the proliferation of nothing but differences, differences without any hierarchical order, differences without fixed identities or biological archetypes, differences generated for their own sake and evaluated only through social and natural contingency. Darwin's work seems to be readily understood through the concept of radical difference even more directly than through the creation of a norm that measures the possibilities of equality. Darwin's understanding of the production of variation is fundamentally embedded in his understanding of sexual selection. Sexual selection proliferates differences as asexual reproduction cannot. Sexual selection must be understood, as Darwin himself does, as the creation of differences without clear models, without pre-given boundaries, differences that have value in themselves for the range of variations they bring to natural (and sexual) selection to select from.

Moreover, Darwin's work on the world of nature is not drastically divisible from his analysis of culture. Nature and culture are not two separate domains, the cultural growing out of the seeds of the natural or somehow moving beyond the natural, superseding it. Rather, the cultural is always already natural, that which the natural enables and that which at least some of the forces of the natural confirm and validate, in the long run, inexorably. Darwin has demonstrated that culture resides within rather than beyond nature: that the most elaborate and complex creatures – ranging from the highly socially stratified insects to various social collectives of primates – complexify nature to create social networks and interrelations that are themselves regulated by natural and sexual selection, by necessity and preference. As is all life on earth, from the most humble and elementary protozoa to the most complex forms of social organization. Darwin has shown that difference is the engine of the natural and its smaller subset, the cultural.

And if Irigaray's work can be understood not only in the sociocultural and linguistic terms in which it has been cast by structuralist and poststructuralist concepts of subjectivity produced only in language but also in terms of the biology of lived bodies, an open-ended and dynamic sense of sexually differentiated bodies that are always in the process of transmutation, then she would not need to account for a transition between nature and culture which figures so centrally in her more recent writings, or to use Hegelian models of the dialectical transformation of and movement beyond nature in order to explain how social and cultural life is possible.

Culture is *not* the movement away from nature, its supersession or transformation as is implied in the paradigms of historical change elaborated in both Hegelianism and in Marxism – or from within phenomenology more generally – but the complication of nature, the functioning of the same broad principles to regulate social and cultural relations as structure natural relations. Irigaray's account of sexual difference as that which has been elided by patriarchal cultures but is nevertheless their unacknowledged condition and which must have its day is only confirmed and strengthened through Darwin's understanding of the pervasive and productive role of sexual selection in the proliferation of differences in nature. It is the force of the natural that insists on sexual difference, and that is a kind of assurance of its return in culture in spite of any forces for its repression (the promise of feminism itself).

Irigaray's work has been painstakingly defended as cultural and political rather than a natural or anatomical analysis. But if a more complex and nuanced account of nature is developed – such as Darwin himself, rather than his followers and interpreters, provides – then concepts of nature need not be tied to essentialist or naturalist frameworks. Nature is dynamized, rendered historical, subject to dramatic change. The human itself is in the process of continual transformation, even beyond itself. Sexual difference remains the most creative and powerful means by which this transformation is brought about. Sexual difference is the means by which the natural cultivates culture rather than culture cultivates and moves beyond nature to create a second nature, or social law in place of nature. We do not leave nature behind, we do not surround ourselves with culture in order to protect ourselves from nature, for culture, cultures in their multiplicity, are themselves complex forms of variation of natural forces, both human and animal.

If Irigaray's work is interpreted through the work of Darwin rather than Hegel, and if Darwin's work is interpreted through the work of Irigaray rather than John Stuart Mill and the politics of egalitarianism, then each is modified through the influence of the other. Each is able to displace the concepts that limit or inadequately frame the implications of their respective positions. Darwin's work is opened to a non-reductive politics of (racial, sexual, ethnic, class) difference, and made relevant to a political analysis of cultural relations. Irigaray's work is strengthened, fortified through the incontestable place of sexual difference as a natural force that is culturally registered, which makes it clear that sexual difference is not just one social difference among many but that form of difference that makes all other lived differences possible, the engine of all lived differences.

Sexual difference is ineliminable, the force that proliferates all social as well as natural relations. Sexual selection refers to the possible erotic relations and encounters between sexes (whether within one sex, between two, or across a number of sexes), sexes themselves separated by difference. Sexual difference is made more and more visible and perceptible over time as the sexes diverge further from each other through the operations of sexual selection. Sexual selection is how sexual difference transforms itself, intensifies itself, selects the most attractive, noticeably forming new ideals, new types of body, qualities and activities. Sexual selection enhances sexual difference; sexual difference proliferates and varies itself through sexual selection. While different, they operate hand in hand to complexify social and natural life and to divide and increasingly differentiate populations. Irigaray and Darwin have each, in separation from the other, come to a point of commonness in which different bodies, divided along the lines of sex, become the means for new natural and cultural relations, the road to new forms of politics and new forms of life.

notes

A shorter, different version of this paper has been published as "Sexual Difference as Sexual Selection: Irigarayan Reflections on Darwin" in *Becoming Undone: Darwinian Reflections on Life, Politics and Art* (Durham, NC: Duke UP, 2011) 143–68.

1 Luce Irigaray, "Sexual Difference" in *An Ethics of Sexual Difference*, trans. Gillian C. Gill and Carolyn Burke (Ithaca, NY: Cornell UP, 1993) 5.

2 Irigaray has always been suspicious of the attempts, primarily in the work of Derrida and Deleuze (and Guattari), to elaborate a politics of "becoming woman" in which sexual difference, while being abstractly recognized, is nonetheless continually undermined by men's attempts to "become woman" without recognizing that such a becoming woman is a fantasy while it functions through the everyday operations of a male morphology. She urges men to cease becoming women and to begin becoming a new kind of man:

> As far as I am concerned, "becoming woman" or "becoming a woman" corresponds to cultivating my own identity, the identity which is mine by birth. For Deleuze, it amounts to becoming what he is not by birth. If I appeal to a return to nature, to the body – that is, to values that our Western culture has scorned – Deleuze acts in the opposite way: according to him it would be possible and suitable to become someone or something which is without relation to my original and material belonging. How could this be possible above all from the part of a man with respect to becoming woman? Putting on the stereotypes concerning femininity? Deleuze would want to become the woman who Simone de Beauvoir did not want to become? (Luce Irigaray, "Becoming Woman" in *Conversations*, trans. and eds. Stephen Pluháček *et al.* (London: Continuum, 2008) 79)

3 See Margaret Whitford, *Luce Irigaray. Philosophy in the Feminine* (London: Routledge, 1990); Carolyne Burke, Naomi Schor, and Margaret Whitford, eds., *Engaging with Irigaray* (New York: Columbia UP, 1994); Ellen Mortensen, *Touching Thought: Ontology and Sexual Difference* (Lanham, MD: Lexington, 2004).

4 "Whether through collective psychosis or cynicism, sexual difference, which constitutes the most basic human reality, is treated like an almost non-existent problem" (Luce Irigaray, "Preface" in *Thinking the Difference: For a Peaceful Revolution*, trans. Karen Montin (London: Routledge, 1994) ix).

5 This is the claim that Irigaray elaborates in "Thinking Life as a Relation" in *Conversations* 5.

6 For more detail, see Luce Irigaray, "Towards a Sharing of Speech" in *Key Writings* (London: Continuum, 2004) 77.

7 Irigaray makes this claim in "The Time of Difference" in *Why Different? A Culture of Two Subjects. Interviews with Luce Irigaray*, ed. Sylvère Lotringer (New York: Semiotext(e)) 96.

8

> Sexuate difference is the most basic and the most universal difference. It is also the difference which operates, or ought to operate, each time, in the connection between nature and culture for everyone. This connection is specific to girl and woman in comparison with boy and man. (Irigaray, "Becoming Woman" in *Conversations* 77)

9 As Irigaray argues in "The Redemption of Women" in *Key Writings* 157.

10 Irigaray claims that sexual difference, while it does not cause or explain all other social differences, provides a necessary mode of engagement which has profound implications for how all social minorities are understood and treated. See "Thinking Life as a Relation" in *Conversations* 18.

11 Ibid. 5.

12 Luce Irigaray, "Human Nature is Two" in *I Love to You: Sketch of a Possible Felicity in History*, trans. Alison Martin (New York: Routledge, 1996) 35.

13 Ibid. 37.

14 Ibid. 39.

15 Luce Irigaray, "The Culture of Difference" in *Je, Tu, Nous: Toward a Culture of Difference*, trans. Alison Martin (New York: Routledge, 1993) 46.

16 Luce Irigaray, "Different but United through a New Alliance" in *Why Different?* 118–19.

17 Luce Irigaray, "A Chance to Live" in *Sexes and Genealogies*, trans. G.C. Gill (New York: Columbia UP, 1993) 16. Subsequent page numbers are given in the text.

18 Luce Irigaray, "Donning a Civil Identity" in *I Love to You* 50–51.

19 Luce Irigaray, "The Universal as Mediation" in *Sexes and Genealogies* 137.

20 Irigaray, "Sexual Difference as Universal" in *I Love to You* 47.

21 This is, in part, the object of investigation of Timothy Lenoir's book *The Strategy of Life: Teleology and Mechanics in Nineteenth-Century Biology* (Chicago: U of Chicago P, 1989).

22 Irigaray, "Donning a Civil Identity" in *I Love to You* 50.

23 See Patricia Gowaty, ed., *Feminism and Evolutionary Biology: Boundaries, Intersections, and Frontiers* (New York: Chapman, 1997), especially the chapters by Helena Cronin, Anne Fausto-Sterling, Margo Wilson, Marlene Zuk, Victoria Sork, and Zuleyma Tang-Martinez. See also Griet Vandermassen, *Who's Afraid of Charles Darwin? Debating Feminism and Evolutionary Theory* (New York: Rowman, 2005).

24 This claim is now most closely associated with Richard Dawkins' book of the same name – *The Selfish Gene*, 30th anniversary ed. (New York: Oxford UP, 2006) – but it is a pervasive assumption within evolutionary thought; see the work of Daniel Dennett and E.O. Wilson.

25 As Dawkins claims, the animal can be reduced to its sex cells and its sex-cells can be largely, indeed solely, explained in terms of the size and quantity of gametes:

> there is one fundamental feature of the sexes which can be used to label males as males, and females as females, throughout animals and plants. This is that the sex cells or "gametes" of males are much smaller and more numerous than the gametes of females [. . .] it is possible to interpret all the other differences between the sexes as stemming from this one basic difference. (Dawkins, *Selfish Gene* 141)

26 Dennett makes explicit the kind of algorithmic reduction of evolution to a step-by-step process, a position I have critiqued in *The Nick of*

Time: Politics, Evolution, and the Untimely (Durham, NC: Duke UP, 2004) 48ff.

27 Helena Cronin asserts that biology provides or should provide an account of human nature, one that is of relevance in the work of policy- and law-makers who address this given nature in a variety of forms to transform behavior enacted on its basis: "All policy-making should incorporate an understanding of human nature, and that means both female and male nature" (Helena Cronin, "Getting Human Nature Right" in *Science at the Edge*, ed. John Brockman (London: Weidenfeld, 2004) 61).

28 In, for example, the more postmodern understanding of the problem of essentialism in the writings of Anne Fausto-Sterling, "Feminism and Behavioral Evolution: A Taxonomy" in *Feminism and Evolutionary Biology* 47.

29 Charles Darwin, *The Descent of Man, and Selection in Relation to Sex* (Princeton: Princeton UP, 1981) I: 94.

30 It must also be noted that, in spite of well-intended commitment to a broad egalitarianism, his understanding of the relations between men and women is in fact quite ambivalent. He affirms in certain places that women are less intellectual, less detached, more sympathetic and warm than men. At times he affirms women's social qualities as if they were biological qualities while at other times he seems to acknowledge that social pressures exert a considerable force in transforming character traits and personal abilities. See, for example, Darwin, *Descent of Man* II: 326–27.

31 Ibid. 366–67.

32 Darwin argues that it is the social treatment of women that requires transformation if women are to attain the pre-eminence of some men (ibid. 329).

33 Karl Marx qtd in Howard E. Gruber and Paul H. Barrett, *Darwin on Man: A Psychological Study of Scientific Creativity* (New York: Dutton, 1974) 71.

34 This is Darwin's first published reference to sexual selection. Charles Darwin, *On The Origin of Species by Means of Transmutation, or Preservation of Favoured Races in the Struggle for Life* (New York: Modern Library, 1998) 117.

35 Darwin, *Descent of Man* II: 254.

36 Darwin, *Origin of Species* 118.

37 See especially Darwin's claims in *Descent of Man* II: 256–58. Subsequent page numbers are given in the text.

38

> Thus the more vigorous females, which are the first to breed, will have the choice of many males; and though they may not always select the strongest or best armed, they will select those which are vigorous and well armed, and in other respects the most attractive. (262)

39 Darwin, *Descent of Man* I: 207–11.

40 Charles Darwin and Alfred Russel Wallace, *Evolution by Natural Selection* (Cambridge: Cambridge UP, 1958) 263.

41 Michael T. Ghiselin, *The Triumph of the Darwinian Method* (Berkeley: U of California P, 1969) 219.

42 Charles Darwin, *A Monograph of the Sub-Class Cirripedia – With Figures of all the Species* (London: FQ, 2010).

43 Ibid. 210–11.

44 Ibid. 207. "In the mammalian class the males possess in their vesiculæ prostraticæ rudiments of a uterus with the adjacent passage; they bear also rudiments of mammæ, and some male marsupials have rudiments of a marsupial sack" (ibid. 208).

45 Ibid.

46 In a letter to his friend Hooker, Darwin explains his discovery of the emergence of maleness in a particular barnacle species:

> I have lately got a bisexual cirripede, the male being microscopically small and parasitic within the sack of the female. I tell you of this to boast of my species theory, for the nearest closely allied genus to it is, as usual, hermaphrodite, but I had observed some minute parasites adhering to it, and these parasites I now can show are supplemental males, the male organs in the hermaphrodite being unusually small, though perfect and containing zoosperms: so we have almost a polygamous animal, simple females alone being wanting. I never should have made this out, had not my species theory convinced me, that an hermaphrodite species must pass into a bisexual species by insensibly

> small stages; and here we have it, for the male organs in the hermaphrodite are beginning to fail, and independent males ready formed. (Darwin qtd in Ghiselin, *Triumph of the Darwinian Method* 115)

47 Rebecca Stott, *Darwin and the Barnacle* (New York: Norton, 2003) 85.

48 Darwin qtd in ibid. 100.

49

> The whole [male] animal is reduced to an envelope [. . .] containing the testes, vesicula, & penis. In male *Ibla*, we have hardly any cirri or thorax; in some male Scalpellums no mouth [. . .] I believe that males occur on every female; in one case I found 12 males & two pupae on point of metamorphosis permanently attached by cement to one female! (Darwin, letter to Hancock qtd in ibid. 213)

50 As Darwin explains, in a letter to Henslow:

> But here comes the odd fact, the male or sometimes two males, at the instant they cease being locomotive larvae become parasitic within the sack of the female, & thus fixed & half embedded in the flesh of their wives they pass their whole lives & can never move again. Is it not strange that nature should have made this one genus unisexual & yet have fixed the males on the outside of the females. (Qtd in ibid. 101)

51

> He who admits the principle of sexual selection will be led to the remarkable conclusion that the cerebral system not only regulates most of the existing functions of the body, but has indirectly influenced the progressive development of various bodily structures and of certain mental qualities. (Darwin, *Descent of Man* II: 402)

52 In, for example, Irigaray, "Introducing: Love between Us" in *I Love to You* 34, 38, where she discusses plants and flowers; and Luce Irigaray, "Animal Compassion" in *Animal Philosophy: Ethics and Identity*, eds. Peter Atterton and Mathew Calarco (London: Continuum, 2004) 195–201.

Life . . . is that which is capable of error.
Michel Foucault, "Life: Experience and Science" in Aesthetics, Method, and Epistemology *476*

[T]here is no sex. There is . . . sex that oppresses and sex that is oppressed. It is oppression that creates sex and not the contrary.
Monique Wittig, "The Category of Sex" in The Straight Mind *2*

What are humans, if not a species capable of relationships that are not subject to instinct?
Luce Irigaray, "Towards a Sharing of Speech" in Key Writings *77–78*

cesare casarino

SEXUAL DIFFERENCE BEYOND LIFE AND DEATH

or, feminism and the biopolitical turn

A specter is haunting philosophy – the specter of life. Whence such a specter? I will begin to answer this question by retelling a familiar story.

By now, it is somewhat of a cliché – though by no means an entirely incorrect or ineffectual one – to describe twentieth-century thought (including philosophy) as having been characterized and defined by a linguistic turn, namely, by a manifold endeavor to reconceptualize everything from the standpoint of language, by a systematic attempt to posit language as the crucial concept for rethinking both being and beings. Put differently, in the twentieth century language was reconceptualized as a principally ontological category, whereas earlier it had constituted almost always an exclusively epistemological category; indeed, for many, language became the ontological category par excellence.[1] In this sense, Martin Heidegger's famous words, in his 1947 "Letter on Humanism," may stand as an appropriate emblem of much twentieth-century thought: "Language is the house of Being. In its home man dwells. Those who think and those who create with words are the guardians of this home."[2] One wonders, however, who is doing all the housework, who is the housekeeper, in this home in which "man dwells." One wonders, in other words, what constituted the (ontic and ontological) condition of possibility for thinking of language in such a way in the first place. One wonders whether such a condition of possibility was appropriately acknowledged and compensated for services rendered, or whether it remained unacknowledged instead, and was exploited, repressed, or foreclosed *tout court*. One wonders whether such a condition of possibility was constituted by the body of sexual difference (where "body of sexual difference" is to be understood as a strictly tautological locution, since, as Luce Irigaray taught us and as Elizabeth Grosz reminds us, sexual difference is always already incorporated and the body is always already sexually differentiated). One wonders, above all, what difference there is between language and being – and, indeed, whether this difference is nothing other than sexual difference itself.

Even though the story of the linguistic turn is well known, I believe it has become necessary to relate it all over again as the story of a compelling attempt, of a beautiful failure: this was at once the attempt and the failure to produce an ontology that would dare to think the gaping abyss and relating non-relation between language and being without, however, closing it, filling it, erasing it – an ontology that would dare to conceive such a chasm as chiasm and hence to think effectively the being of language as the language of being. In short, this was the attempt and the failure to think the immanence of language and being adequately. A stupendously productive ontological endeavor . . .

Arguably, the most significant product and most conspicuous index of such an attempted yet failed ontological endeavor is our current and ongoing obsession with life: life may turn out to be the enduring, obstinate inheritance bequeathed to us by the linguistic turn. Ironically, one of the earliest and most influential thinkers of the linguistic turn might have been also one of the first to put his finger precisely on this emergent problem and to sense with rare prescience the eventual outcome of the ontological adventures of language. I am thinking of Ludwig Wittgenstein and especially of that moment in his *Philosophical Investigations* when he asserted forcefully and in unambiguous terms that "[T]o imagine a language is to imagine a form of life."[3] Wittgenstein understood well – and often much better than many of his progeny – that language is both words and bodies, that language is always incorporated in living forms. Could it be that the linguistic turn had been driven by a rarely acknowledged, seldom understood, and poorly articulated desire to grasp life and its forms all along, a desire to conceive notions of life different from the various nineteenth-century positivist, organicist, or vitalist ones? I am suggesting that both the success of the linguistic turn (namely, to produce language as the ontological concept par excellence) and its failure (namely, its inability to think the immanence of language and being adequately, thereby re-transcendentalizing either language or being or both) shared in one and the same condition of possibility: the foreclosure of the concept of life and of its bodies or forms, the *Verwerfung* of life itself. Such a *Verwerfung* is a very dangerous language game indeed: foreclosure of life turns the real into a hallucinogenic swamp from whose hazy vapors spectral chimeras of life emerge and return to us zombie-like ready to haunt, possess, and devour our discourse (philosophical and otherwise). On the indistinct horizon of this swamp, it is very difficult at times to discern life itself from its own transcendent delirium, to distinguish forms of life from their own essentialist mirages and simulacra. Is our contemporary concern with the question of life a concerted effort to address and redress the various (either partial or total) foreclosures of the concept of life that characterized the linguistic turn? Or is it yet another psychotic symptom of a discourse of life gone awry, yet another ill-begotten specter of an enduring foreclosure? And could it be both? I ask such questions precisely as someone who finds it more than ever necessary and urgent to think and to write about life today.[4]

Be that as it may, what is certain is that during the last two decades of the twentieth century the linguistic turn gradually gave way to what I would like to call *the biopolitical turn*. During this period, for better and for worse, a shift started taking place from "language" to "life" as the guiding category for ontological projects of all sorts. Such a shift was not at all a fortuitous happenstance. The intimate and symbiotic connection between the linguistic turn and the biopolitical turn, for example, finds nowhere stronger evidence and confirmation in the fact that a crucial protagonist of the former was also the thinker who ushered in the latter and who famously coined the term "biopolitics" in 1974, namely, Michel Foucault.[5] Foucault arguably spent several decades articulating the relations between language and being, between words and things, before finding it necessary and inescapable somehow to turn his thought to the ways in which in modernity life had been at once produced, captured, and mortified by sovereign power, capitalist exchange relations, and techno-scientific discourse at one fell swoop. In Foucault's wake, the philosophical critique of biopolitics, in its various and often divergent forms, has unfolded also as an attempt to reclaim life at once from state, capital, and techno-science, as an attempt to think life otherwise and to produce new and indeed more vital forms of life.

I understand Elizabeth Grosz's influential and multifaceted work as having been at the forefront of this philosophical project (consisting at once of a critique of biopolitics as well as of a reclaiming and a reconceptualization of life) for the last two decades – though, to be sure, her work is by no means limited to and definable in terms of such a project alone. In particular, her work has been especially enmeshed in this philosophical project since the publication of *The Nick of Time: Politics, Evolution, and the Untimely* in 2004 and *Time Travels: Feminism, Nature, Power* in 2005 – two books whose main concerns are reflected and elaborated further in the strikingly unorthodox and thought-provoking essay on Luce Irigaray and Charles Darwin included in this special issue of *Angelaki.* Grosz opens this essay with a self-declaredly "bold conjecture": "perhaps Irigaray's work on sexual difference might find unexpected confirmation from Darwin's account of sexual selection."[6] If I situate such a "bold conjecture" – as well as her work more generally – within the philosophical genealogies I have articulated above, that is not merely because life is virtually ubiquitous in it. Though "life" is indeed a highly recurrent term in Grosz's paper on Irigaray and Darwin, and though references to the "biological" in the widest sense of that term abound throughout, what is significant here is surely neither a matter of quantity nor a matter of terminology per se. In other words, life looms large on this essay not only terminologically but also and more importantly conceptually. The entire paper, after all, ends with a poignant call for "new forms of life," and I assume that altogether new concepts of life would have to be attendant to and constitutive of such new forms.

Let me be clear: I presume neither to do justice to Grosz's entire work nor to engage with and to evaluate the validity of the specific arguments regarding Irigaray and Darwin that she articulates in this essay. Even though ultimately I would answer the question "Can biology provide feminist thought with the conceptual resources by which to understand sexual difference?" largely dissentingly, here I am more concerned with pondering why one would want to ask such a question in the first place. Even though in the end I am not entirely sure that Irigaray and Darwin need one another, and that, as Grosz contends, "by cross-fertilizing their quite different contributions, the work of each can be sharpened, made more conceptually incisive, more broadly relevant," I am not concerned with assessing and contesting such undeniably daring, intriguing, and provocative claims here. I intend, rather, to reflect more generally on what it means to hypothesize and advocate "a kind of cross-mutation, an unexpected evolutionary effect of the coupling of the two, Darwin and Irigaray."[7] I want to reflect on what it means to want to hybridize Irigaray and Darwin today, on what might impel one to twist and turn and intertwine these two thinkers into a veritable Möbius strip, on what is at stake in this most improbable, most startling, and most perverse of couplings.[8] As I will try to show in what follows, whatever else is also at stake in such a coupling, Grosz's bold conjecture involves and presupposes precisely an attempt at once to reconceptualize and reclaim life itself for the feminism of difference as well as to bring such a feminism to bear on the current and ongoing, discursive and non-discursive redefinitions of life. This is why it has seemed appropriate to me to frame this inquiry into her wager in terms of the passage from the linguistic turn to the biopolitical turn.

As Grosz navigates deftly in that zone of indistinction between the Scylla of natural life in Darwin and the Charybdis of cultural life in Irigaray – each of which she shows to be far from internally homogeneous or self-identical – it seems that what is at stake here is a concept of life that would displace radically such a dialectical binary once and for all by rethinking it as an irreparably asymmetrical dyad instead. Grosz writes: "There can be no direct reconciliation between Irigaray and Darwin to the extent that each is implicated in quite different projects, Irigaray in philosophy, and addressing only the human, Darwin in biology and addressing primarily the animal."[9] If no dialectical synthesis is possible here, and if hence this encounter must remain oblique and aslant, that is because there is neither symmetry nor opposition between the so-called cultural and the so-called natural to begin with. Grosz writes: "Darwin has demonstrated that culture resides within rather than beyond nature."[10] Herein the Spinozist inspiration and aspiration of Grosz's project become apparent, since Baruch Spinoza – arguably an immensely influential and important thinker for Grosz – had asserted already in the seventeenth century that the position of culture with respect to nature is precisely *not* one of *imperium in imperio*, of an empire within an empire, namely, of a self-enclosed, autonomous, and transcendent enclave beyond and outside of which there lies a world of natural life effected by an entirely different order of causality and functioning according to an entirely different set of rules and regulations.[11]

What are, then, the singular cause as well as the shared common rules and regulations that in effect constitute the relation between nature and culture as a topological relation? For Grosz, cause turns out to be sexual difference, while the set of rules and regulations cutting across all of nature – including culture – consist of sexual selection. Grosz writes that Darwin's inquiry into sexual selection "has shown that difference is the engine" of "all life on earth," "of the natural and its smaller subset, the cultural." Moreover, on the basis of her cross-fertilization of Darwin's insight with Irigaray, Grosz proceeds to argue that, on the one hand, "[s]exual difference is ineliminable, the force that proliferates all social as well as natural relations," and, on the other, "[s]exual selection is how sexual difference transforms itself, intensifies itself."[12] In short, Grosz's Spinozist hybridization of Darwin and Irigaray leads her to articulate the relation between nature and culture as a relation of immanence, to understand sexual difference as immanent causality, and to argue that "all life on earth" – that is, all cultural as well as natural relations and forms – is caused and determined by sexual difference just as it re-produces and re-determines sexual difference through sexual selection. In a surprising turn of events – which undoubtedly would have Spinoza turning in his grave, and about which I cannot imagine Irigaray herself being too pleased either – Grosz posits sexual difference here as nothing other than Spinoza's *Deus sive Natura*, God or Nature, namely, immanent cause of itself and of all modes of being, of all forms of life. This is perhaps why Irigaray – even though she is not particularly partial or sympathetic to this unquestionably misogynistic thinker – when writing about Spinoza's understanding of God in her work *An Ethics of Sexual Difference* nonetheless does concede intriguingly that: "it is necessary to go through the question of God every time the sexual act comes under consideration."[13] In short, immanence is always at stake when attempting to think not only sexual difference and sexual selection but also sexual relation – namely, the relations established however fleetingly in the sexual act – a relation which itself, thus, ought to be understood as a topological relation in its own right.

I will return to these matters, and to the question of sexual selection in particular. For the moment, I would like to draw some provisional conclusions and to refine my initial characterization of Grosz's work. It is not simply the case that her work needs to be understood, on the one hand, as part and parcel of the contemporary philosophical discourses on biopolitics, and, on the other, as not confined to and by these discourses; more importantly, it is also the case that her work constitutes a pointed intervention in the discourses on biopolitics, an intervention that intends to map the as yet uncharted crossroads where biopolitics and patriarchy meet, an intervention that intends to make the concept of sexual difference inescapable for the thought of biopolitics (that is, inescapable both for the critique of biopolitics and for the reconceptualization of life that such a critique ought to involve). In a sense, this is the other underlying and more implicit hybridization called for in Grosz's essay: in effect, this essay as well as the two aforementioned books from which it originates constitute pioneering works in what one day might well be known as biopolitical feminism.

Any attempt to reclaim and to redefine life, and hence any thought of biopolitics, however, does have its dangers. If the linguistic turn was founded on a foreclosure of the concept of life, the biopolitical turn, conversely, in its attempt to counter such a foreclosure, might well run the reactive risk of foreclosing language and reifying, fetishizing, and hence re-transcendentalizing life once again. Such a risk is a consequence of the fact that there where one attempts to reconceptualize life, one operates at the least in the vicinity – if not at times within the same set of procedures – of the biopolitical capture of life. In short, the very real risk of reinstating exactly those vitalist essentialisms to which the linguistic turn had understood itself as a corrective is itself a product of that power which has been and still is involved in producing, re-producing, controlling, and harnessing life: the reification and fetishization of life as commodity (conceptual and otherwise) needs to be understood as a technology of biopower in its own right. (And, indeed, understood in the context of what has been termed and theorized as contemporary biocapital, doesn't the current and ongoing resurgence of vitalism as a philosophical discourse constitute – besides whatever else it may also constitute – an eminent instance of commodity fetishism, that is, of fetishization of that most precious of commodities which is labor power?)[14] If Wittgenstein may have sensed the risks inherent in the ontological investigation of language, several thinkers of biopolitics are acutely aware of the risk involved when handling that most elusive, hazardous, and overdetermined of signifiers, namely, life. Regarding biopower, for example, Foucault writes:

> Against this power that was still new in the nineteenth century . . . the forces that resisted relied for support on the very thing it invested, that is, on life and man as a living being . . . Life as a political object was in a sense taken at face value and turned back against the system that was bent on controlling it.[15]

Commenting on Foucault, Gilles Deleuze writes: "Life becomes resistance to power when power takes life as its object . . . [T]he two operations belong to the same horizon."[16] Giorgio Agamben, commenting on both these passages, concludes: "Like Foucault, Deleuze is perfectly conscious of the fact that any thought that considers life shares its object with power and must incessantly confront power's strategies."[17] All of the above means, among other things, that life itself does not pre-exist its own capture: it is only because of a series of processes that go by the name of biopolitics that we can at all speak of life – let alone be concerned with reclaiming and redefining it. This is why we ought to say of life exactly what Monique Wittig says of sex in the epigraph to this essay: there is no life; there is life that captures and life that is captured; it is the set of power relations consisting of capture (namely, biopolitics) that produce life rather than the contrary. There is no life – yet there are plenty of life effects.

Unlike Wittig, however, I will not draw the analogous conclusions that hence the category of life is not an ontological category and that we should just abolish it (as if we could do so by fiat even if we wanted to).[18] Life and sex may well share a common political genealogy as well as a common destiny. Dire risks indeed are those facing anyone who wants to think life and sex today, especially if in terms of each other: the manifold risks of transcendence . . . And yet – as Deleuze, among others, points out – transcendence is always a product of immanence: transcendence is the very real illusion one always may encounter and the very real risk one always runs whenever attempting to construct a plane of immanence and to think such a plane of immanence as the unthought within thought, as that which must be thought and yet cannot be thought, and hence as that which one can only attempt to show that it is there, unthinkable and unrepresentable, in its effects.[19] That is to say that such risks do not constitute a good reason not to think, not to write, or not to talk about life and sex; they constitute, rather, a good reason simply to remind ourselves of the word Spinoza had engraved on his signet ring, with which he sealed all his letters: *caute*, cautiously.

Let me end and seal this essay, thus, with a (hopefully cautious) series of questions (which may yield nonetheless a quite possibly reckless, intransigent, contentious answer). Nowhere are life and sex more intertwined perhaps than in Sigmund Freud's theory of the drives and in his concept of the death drive – nowhere, that is, outside of Jacques Lacan's reformulation of this theory and of this concept.[20] (And let me add that I understand Lacan's reformulation in essence as positing the concept of the death drive simply as *drive*, in the sense that so-called life and so-called death may be said at most to constitute two fundamental aspects of the drive, as well as in the sense that any drive whatsoever, *mutatis mutandis*, functions in the exact same ways as Freud's death drive.) Drive – which, of course, is precisely not instinct – is the engine of human life and is always sexual. Alien to any homeostatic tendency, and yet always embedded there and ready to disrupt any such tendency, drive is productive process par excellence. It produces and re-produces human life according to three main modalities: production of excess, production of repetition, and production of death. In short, drive is always excessive, always repetitive, and always destructive and potentially fatal (where "destructive" and "fatal" are intended strictly according to Spinoza's definition of death as no more and no less than essential transformation or radical metamorphosis).[21] If I invoke drive in this context, that is because the similarities between drive and Grosz's reformulation of sexual selection are striking, to say the least. Indeed, there are passages in her essay in which it seems as if sexual selection is to drive precisely what natural selection is to instinct. Grosz writes:

> Darwin carefully distinguishes sexual from natural selection. While natural selection privileges sexual reproduction over the forms of asexual reproduction that characterize some forms of life (bacterial, viral, protozoan), sexual selection is a principle irreducible to natural selection. Natural selection is always ultimately directed by the struggle for existence, the struggle to survive. It is oriented to factors regulating life and death . . . [S]exual selection is not only a separate principle from natural selection, for it could have never been deduced from a knowledge of natural selection; sexual selection attenuates and problematizes the criteria by which natural selection operates and substitutes its own, sometimes contrary, principles of taste, attractiveness or desire. Sexual selection not only complicates natural selection; it has the potential to imperil life, to render various activities or qualities more noticeable, more obvious and as vulnerable to attracting predators as much as potential sexual partners.[22]

Unlike Darwin and unlike Irigaray, Grosz rethinks both the animal and the human, both nature and culture. Moreover, as I have stated above, she wants to rethink them not as opposed and transcendent with respect to one another but as absolutely immanent – and this is a project that I find to be necessary, urgent, and, indeed, paramount. And yet, if it is the case that sexual selection is that which in all forms of life – natural as well as cultural – at once is caused by and re-determines sexual difference by proliferating it, intensifying it, and repositioning it on ever higher planes of complexity, surely it is also the case that all forms of life re-determine sexual difference differently: this is precisely where the distinction between instinct and drive may come into play, in the sense that drive may constitute the *differentia specifica* of the human re-determination of sexual difference. In short, here, at the spellbound quadrivium of instinct, drive, natural selection, and sexual selection, I find that I have lost my way, and that I am facing a possibly insoluble conundrum. Hence the questions (and their arbitrary answer):

If sexual selection is indeed related to drive, are we to infer then that all forms of life are subject not only to instinct but also to drive (thereby potentially erasing or at least blurring the distinction between the concepts of instinct and of drive altogether)? Or, if we still find it useful to retain the distinction between these two concepts, are we to posit it as a relative and quantitative difference, a difference of degree, rather than as an absolute and qualitative difference, a difference in kind? And if somehow we still insist, on the one hand, on retaining the distinction between instinct and drive as difference in kind, and, on the other, on positing both instinct and drive as constitutive of all forms of life, are we then to suppose, perhaps, that forms of life differ only in the ways and in the degrees in which they combine instinct and drive (rather than differing due to the complete absence of drive in some forms and due to the defining presence of drive in some other forms)? Undoubtedly, excess, repetition, and death are not human monopolies. But how exactly are excess, repetition, and death *definitional* of (which is to say also *problematic* for) any form of life other than the human one? And if, rather, sexual selection and natural selection are understood to constitute different aspects of instinct only – its two fundamental aspects, perhaps – how are we to account, then, for the similarities between sexual selection and drive? In short, do Grosz's Irigarayan reflections on Darwin any longer allow or even have any use for a concept of drive that would be distinct from a concept of instinct? Finally, I ask – in the wake of and along with Irigaray – "What are humans, if not a species capable of relationships that are not subject to instinct?" I know no better answer to this poignant question than Irigaray's own irrefragable answer: "That is what we should be, in any case."[23] An answer that is also an injunction and a wager: an injunction that interpellates us as humans even though or precisely because it is acutely cognizant of the real possibility that we may well err and stray from its interpellation; a wager that human life is that which is capable of error (since instinct never errs), and is, thus, an irreparably erratic, errant, ethical form of life.

notes

This essay is a revised version of my response to the lecture which Elizabeth Grosz delivered at the "What is Sexual Difference Now?" symposium, 13 Nov. 2009, at the University of Minnesota, and which constituted the basis for her essay included in this special issue of *Angelaki*. I am very grateful to my friends and colleagues Arun Saldanha and Hoon Song for having invited me to participate in the symposium as a respondent to Grosz, as well as for their enthusiastic comments and repeated encouragements that have proved essential for revising my response into this essay. I am very grateful also to Grosz for having given me the opportunity to ponder such vital matters and hence for having produced much thought in me.

1 Arguing against such a historiographical cliché (namely, against the characterization of twentieth-century thought as having been marked above all by a linguistic turn), Alain Badiou writes around the end of that century:

> Our epoch can be said to have been stamped and signed, in philosophy, by the return of the question of Being... When all is said and done, there is little doubt that the century has been ontological, and that this destiny is far more essential than the "linguistic turn" with which it has been credited. (Alain Badiou, *Deleuze: The Clamor of Being*, trans. Louise Burchill (Minneapolis: U of Minnesota P, 1999) 19)

The reason why, unlike Badiou, I find the cliché of the linguistic turn to constitute a largely adequate assessment of twentieth-century thought is that what he calls "the return of the question of Being" happened to take the route of language, that "Being" returned by traveling along the path of language, and hence that the return of being and the turn to language are in strictly complementary and symbiotic relations to one another: being returned via language and language turned to being at one and the same time. Moreover, as I will suggest shortly, the return to ontology since the beginning of the twentieth century has been characterized by two interrelated yet distinct phases and aspects, only one of which was constituted by the linguistic turn, whereas the other (and ongoing) one is what I refer to as the biopolitical turn.

2 Martin Heidegger, "Letter on Humanism" in *Basic Writings*, ed. David Farrell Krell; trans. Frank A. Capuzzi with J. Glenn Gray (New York: HarperCollins, 1993) 217.

3 Ludwig Wittgenstein, *Philosophical Investigations*, eds. P.M.S. Hacker and Joachim Schulte; trans. G.E.M. Ascombe (Chichester: Wiley-Blackwell, 2009) 11.

4 These questions, in other words, are not rhetorical; they are, rather, questions to which I have not found entirely satisfactory answers yet and which I continue to ask not only others but also

myself. See, for example, "Three Theses on the Life-Image (Deleuze, Cinema, Biopolitics)" in *Releasing the Image: From Literature to New Media*, eds. Jacques Khalip and Robert Mitchell (Stanford: Stanford UP, 2011) 156–67.

5 Before his 1978–79 lectures at the Collège de France on the birth of biopolitics, Foucault mentions the term biopolitics for the first time in a lecture titled "The Birth of Social Medicine," delivered in Rio de Janeiro in 1974. He returns to this term and develops it further in 1976, in that year's lectures at the Collège de France as well as in the last chapter of his *History of Sexuality, Volume I* (New York: Vintage, 1990). Michel Foucault, "The Birth of Social Medicine" in *Essential Works of Foucault, 1954–1984: Power*, ed. James D. Faubion; trans. Robert Hurley and others (New York: New, 2000) 134–56; idem, *"Society Must be Defended": Lectures at the Collège de France 1975–1976*, trans. David Macey (New York: Picador, 2003), especially 239–72.

6 Elizabeth Grosz, "Sexual Difference as Sexual Selection: Irigarayan Reflections on Darwin," paper presented at the "What is Sexual Difference Now? A Conversation with Joan Copjec and Elizabeth Grosz" symposium, University of Minnesota, 13 Nov. 2009.

7 Ibid.

8 I am invoking the Möbius strip model advisedly, since this is a model that Grosz at once deploys and critiques in effective ways in the closing pages of what is arguably one of her most important books, namely, *Volatile Bodies: Toward a Corporeal Feminism* (Bloomington: Indiana UP, 1994) 209–10. It goes without saying that my characterization of the Irigaray–Darwin coupling as "most perverse" is informed by the critical investigation, re-elaboration, and reclaiming of the question of perversion by Grosz and others (such as Teresa de Lauretis and Judith Butler) rather than by the always patriarchal (as well as almost invariably sanctimonious and censorious) deployment of perversion in Freudian psychoanalysis, if not also in Sigmund Freud's works; for Grosz's engagement with the question of perversion, see at least the essays comprising the third and final section of her *Space, Time, and Perversion* (New York: Routledge, 1995) 141–227.

9 Grosz, "Sexual Difference as Sexual Selection."

10 Ibid.

11 Spinoza makes this argument both implicitly and explicitly on numerous occasions throughout his works, but see especially the opening paragraph of the Preface to Part Three of his *Ethics*, in which the expression *imperium in imperio* (empire within an empire) occurs. Spinoza, *Ethics*, ed. and trans. G.H.R. Parkinson (Oxford: Oxford UP, 2000) 163.

12 Ibid.

13 Luce Irigaray, *An Ethics of Sexual Difference*, trans. Carolyn Burke and Gillian C. Gill (Ithaca, NY: Cornell UP, 1993) 93.

14 The term and concept of "biocapital" has been articulated admirably by Kaushik Sunder Rajan in his *Biocapital: The Constitution of Postgenomic Life* (Durham, NC: Duke UP, 2006). On these and related matters, however, see also Melinda Cooper's important *Life as Surplus: Biotechnology and Capitalism in the Neoliberal Era* (Seattle: U of Washington P, 2008).

15 Foucault, *History of Sexuality, Volume I* 145.

16 Gilles Deleuze, *Foucault*, trans. Seán Hand (Minneapolis: U of Minnesota P, 1988) 92.

17 Giorgio Agamben, "Absolute Immanence" in *Potentialities*, trans. Daniel Heller-Roazen (Stanford: Stanford UP, 1999) 232.

18 This is in essence the main conclusion reached by Wittig regarding the category of sex in the essay from which the epigraph to my essay is extracted. Monique Wittig, "The Category of Sex" in *The Straight Mind and Other Essays*, trans. Louise Turcotte (Boston: Beacon, 1992) 1–8.

19 Deleuze made this argument in a variety of ways throughout his life, but see especially the chapter (with Guattari) on "The Plane of Immanence" in *What is Philosophy?*, trans. Hugh Tomlinson and Graham Burchell (New York: Columbia UP, 1994) 31–60, especially 59–60.

20 The fundamental text for the theorization of the death drive is, of course, Freud's *Beyond the Pleasure Principle*. For Jacques Lacan's critical re-elaborations of the theory of the drive, see at least his *The Four Fundamental Concepts of Psychoanalysis*, ed. Jacques-Alain Miller; trans. Alan Sheridan (New York: Norton, 1998).

21 For Spinoza's arguments regarding death, see at least the Scholium to Proposition 39 in Part Four of the *Ethics* 256–57.

22 Grosz, "Sexual Difference as Sexual Selection."

23 Luce Irigaray, "Towards a Sharing of Speech" in *Key Writings*, trans. Gail Shwab (London: Continuum, 2004) 78.

charles shepherdson

THE BODY, SEXUALITY, AND SEXUAL DIFFERENCE

A rigorous language, as it arises from sexuality, will not reveal the secret of man's natural being, nor will it express the serenity of anthropological truths, but rather, it will say that he exists without God . . . Thus, at the root of sexuality . . . at the root of this discourse on God which Western culture has maintained for so long . . . a singular experience is shaped: that of transgression. Perhaps one day it will seem as decisive for our culture as the experience of contradiction was at an earlier time for dialectical thought. But in spite of so many scattered signs, the language in which transgression will find its space . . . lies almost entirely in the future.

Michel Foucault, "Preface to Transgression" 30–33

This book examines the problems of a humanistic discipline that has been touted as a science on the model of the medical or natural sciences, and has been organized institutionally as a religion.

Douglas Kirsner, Unfree Associations: Inside Psychoanalytic Institutes *6*

I inventing the "body"

In an early text from his so-called "pre-psychoanalytic writings" titled "Some Points for a Comparative Study of Organic and Hysterical Motor Paralyses" of 1888, Freud provides an important early account of the problem of the "body" in psychoanalysis.[1] This text, originally written in French and deriving from his early work in Paris with Charcot, at a time when Freud was still oriented by his medical training in neurology, nevertheless shows at the same time an important break between psychoanalysis and the biomedical framework of organic medicine. Indeed, while he shows genuine respect for Charcot's work, Freud simultaneously dares to criticize his teacher, and he takes his distance from Charcot precisely on the question of this division between psychoanalytic accounts of the symptom and biological models of disease.

In order to grasp this point, let us recall Freud's argument in this famous text. Its first section distinguishes between two kinds of motor paralysis which, Freud tells us, "clinical neurology recognizes" (160). This is the distinction between "periphero-spinal" and "cerebral" paralysis. This distinction "agrees entirely with the findings of the anatomy of the nervous system," which show that the "motor conductive fibres" fall into two distinct sections. The first runs from the extremities of the body to the spinal cord, while the second runs from the spinal cord to the cerebral cortex. As a result, periphero-spinal paralysis, such as "the facial paralysis in Bell's palsy," affects a discrete portion of the body ("in these disorders," Freud writes, "each muscle – one might say each muscle fibre – can be paralysed individually and in isolation"); whereas cerebral paralysis "is always a disorder that attacks a large portion of the periphery, a limb, a segment of an extremity or a complicated motor apparatus" (161).

To explain this difference, which he takes to be already well established, Freud turns to the structure of the nervous system. The nervous "cords" that lead from the "periphery" of the body to the spine correspond "point by point and element by element" to the extremities that they represent. One can therefore speak of a faithful "correspondence" or "reproduction" in the spine of the extremities that are being "represented" by these cords (all these terms are Freud's). In the case of periphero-spinal paralysis, we thus have a faithful "projection" upon the spinal cord of the periphery of the organism (we will leave it to film theory to explore the difficulties that arise from this quasi-visual form of "projection," a form of representation perhaps not unlike what one finds in fiber-optic cable). Cerebral paralysis, by contrast, does not show such a faithful representation. In this case, the number of fibers running through the spinal cord to the cerebral cortex are not sufficient to maintain a "point by point and element by element" correspondence with the periphery. Instead, the cords of the periphery, when they reach the spine, must be bundled into groups, "which no longer each represent a single element of the periphery, but rather a group of them." Freud emphasizes "a change in arrangement at the connecting point between the two sections of the motor system," and continues: "I argue, therefore, that the reproduction of the periphery in the cortex is no longer a faithful reproduction point by point, that it is no longer a true projection." It is, rather, "a relation by means of what may be termed representative fibres, and for cerebral paralysis I propose the name of *representation paralysis*" (161).

Since this remark may be such tempting territory for proponents of cultural studies, and may seem to open the door to speculation about the "signifier" and "symbolic" representation, let us remember that Freud is still describing the received doctrines of neurology. Nevertheless, the grouping of peripheral cords into bundles at the spine (such that each single cord in the spine "represents" several peripheral cords), can explain why the symptoms of these two types of paralysis will differ. Damage in the former system of nervous cords – periphero-spinal paralysis – will affect an individual muscle fiber, while cerebral paralysis will affect a larger "segment of an extremity or a complicated motor function." The reason for this distinction is anatomical. If one imagines a transparency – the sort of colored diagrams that allow medical textbooks to visualize the nervous system, skeletal system, cardiovascular system, etc., such that each page with its image can be superimposed on the next – one can see that Freud is describing received neurological knowledge, by distinguishing these two systems. The system that runs from periphery (A) to spine (B) is distinct from the system that runs from the spine (B) to the cerebral cortex (C), and consequently, *one can locate the "lesion"* that is responsible for each type of paralysis *in the appropriate part of the nervous system*. This explains why, "in cerebral paralysis of the extremities, it can be observed that the distal segments always suffer more than the proximal ones; for instance the hand is more paralysed than the shoulder" (161). The symptoms of these two types of paralysis differ according to these distinct anatomical sections of the nervous system.

Freud now turns to the problem of hysteria, which "has often been credited with a faculty for *simulating* the most various nervous disorders" (162). And yet, in its symptomatology, hysteria "never simulates periphero-spinal or projection paralysis." Hysterical symptoms always "correspond to representation paralyses or organic cerebral paralyses." They always affect a large segment of the motor apparatus, or a function such as walking. Despite this "simulation" of cerebral paralysis, however, the hysterical symptom must in turn be detached from organic paralyses in several respects.

> In the first place, it is not bound by the rule, which applies regularly to organic cerebral

> paralyses, that the distal segment is always more affected than the proximal one. In hysteria, the shoulder or the thigh may be more paralysed than the hand or the foot. Movements may appear in the fingers while the proximal segment is still absolutely inert . . . in contradiction to the rules of organic cerebral paralysis. (162–63)

As a result, hysterical paralysis may also be regarded as a "representation paralysis," Freud says, "but with *a special kind of representation whose characteristics remain to be discovered*" (163; emphasis mine).

Freud now outlines some differences that distinguish hysterical symptoms from those of organic paralysis. In hysteria one finds "the total abolition of a function," such as the ability to stand up, "while another function performed by the same organs remains intact" (164). One finds total paralysis of a limb while the hand or foot remains functional, in contradiction of the laws of anatomy. One finds hysterical aphasia creating "total aphasia (motor and sensory) for a particular language without in the slightest interfering with the faculty of understanding and articulating another," something which is "unheard-of in organic aphasia" (164). In each case, the hysterical body no longer follows the logic of neurology. These differences lead Freud to the dramatic assertion that "*in its paralyses and other manifestations hysteria behaves as though anatomy did not exist*" (169).

The general importance of these observations for the history of thought cannot be overestimated. For with these observations, Freud brings us to the brink of a new division, separating the *biological organism* from the *hysterical body*. Psychoanalysis as a domain of knowledge will henceforth concern itself with a distinctive field – not only a network of concepts and terminology (jouissance, object a, the ego, the unconscious, etc.) but also a domain of newly emerging objects (sexuality, drive, libido, the body, the symptom, etc.) that have no equivalent in the natural organism.

A clear division thus appears to be established between psychoanalysis and biomedical discourse, but one can nevertheless see in Freud's own formulation the difficulty of maintaining this division, and the peculiar attachment Freud maintains – whether unwittingly or by a more conscious strategy of seeking "legitimacy" for his ideas – to the language of established knowledge. Thus Freud considers the possibility of preserving the discourse of neurology, such that cerebral paralysis (which he calls "representation paralysis") would continue to have some common ground with hysterical paralysis: "it may be maintained that hysterical paralysis is also a representation paralysis," Freud says, and he goes on to sustain this very proposal, but at the same time he insists that the logic of the symptom in hysteria involves "*a special kind of representation whose characteristics remain to be discovered*" (163). The border between psychoanalysis and its disciplinary neighbors is thus clearly and even triumphantly announced, while at the same time its capacity for being blurred remains evident.

As I have argued elsewhere, one sees precisely the same erasure of psychoanalysis when the concept of "sexual difference" is interpreted in terms of the more familiar distinction between (biological) "sex" and (social) "gender." Whether Lacan is denounced for maintaining a version of essentialism by insisting on a rigid "law" of sexual difference that covertly repeats the biological model of binary natural sex, or whether he is celebrated for insisting upon the "symbolic" organization of sexual difference in a manner that supports the arguments of social construction, in either case the fundamental conceptual framework of psychoanalysis has already been lost, translated into the more familiar opposition between "nature" and "culture" inherited from the nineteenth century, and conceptually developed by the discourses of *Naturwissenschaften* and *Geisteswissenschaften* ("natural" sciences and "human" or "historical" sciences).[2] A similar translation takes place when Lacan's claim that "man's desire is the desire of the Other" is interpreted by reference to Hegel, the master–slave dialectic, and the "desire for recognition," as routinely happens in the reception of Lacan. In this case, we are told that desire has no natural object, but rather aims at *the desire of the other*, such that desire is always trapped in a movement of rivalry and identification (mimetic desire) that transcends biological need, and can never be satisfied. Human desire is thus always a "desire for recognition," situated beyond biological need, in an intersubjective relation that inherits the transcendental, spiritual and dialectical movement of spirit, but no longer has any relation to the body, or sexuality, or the symptom, or any of the foundational issues of Freudian theory.[3] Once again, psychoanalysis is made to disappear before our eyes, in favor of a more familiar (but pre-Freudian) interpretive framework. Thus, whether Freud is taken up within contemporary discourses of gender and the social network of symbolic constructions, or whether he is reread in the context of Hegelian dialectic, as revealing a desire exposed in its fundamental "negativity," the result is that psychoanalysis has already been lost, translated into a more recognizable discourse.[4] This is why Foucault, writing in 1963, emphasizes the historical rupture that psychoanalysis represents in the history of thought, noting that the discovery of sexuality, and the experience of transgression which it entails, will one day "seem as decisive for our culture as the experience of contradiction was at an earlier time for dialectical thought," while at the same time noting that "in spite of so many scattered signs, the language in which transgression will find its space... lies almost entirely in the future" (see epigraph). Psychoanalysis, from this perspective, will only emerge in its own theoretical specificity once it has been freed from the inherited discourses of both natural science and social construction.

With this broader horizon in mind, let us return to Freud's text. Freud closes his descriptive account of hysterical symptoms (section IV of his text) with an additional observation that merits attention here. "Hysteria is ignorant of the distribution of the nerves," he says. "It takes the organs in the ordinary, popular sense of *the names they bear*: the leg is the leg as far up as its insertion into the hip, the arm is the upper limb *as it is visible under the clothing*" ("Motor Paralyses" 169; emphasis mine). We face an alternative here that Freud does not address: in being detached from the laws of anatomy, does the hysterical symptom follow the logic of the image, or the logic of the name? Is the symptom organized by the "imaginary body" (the "upper limb as it is *visible under the clothing*"), or does it follow the logic of the signifier, taking the organs "in the ordinary, popular sense of *the names they bear*"? What exactly is the causality of the hysterical symptom, once it has been distinguished from organic damage or disease? Lacan will, of course, develop his account of the imaginary and symbolic with precisely these issues in mind, and in fact it is not a question of choosing, since the symptom in psychoanalysis can be organized in relation to both imaginary and symbolic forms of "representation."

Even in texts that belong to the most triumphantly "linguistic" moment of Lacan's work, such as "The Function and Field of Speech and Language in Psychoanalysis," we see more than one gesture, and more than one explanation. On the one hand, Lacan asserts the primacy of language:

> In order to free the subject's speech, we introduce him into the language of his desire, that is to say, into the *primary language* in which, beyond what he tells us of himself, he is already talking to us unknown to himself, and, in the first place, in the symbols of the symptom.[5]

On the other hand, he observes that in the symptoms of neurosis, which bear on the unconscious rather than the organism, we still find a bodily materialization whose logic needs further exploration:

> speech is driven out of the concrete discourse that orders the subject's consciousness, but it finds its support either in the natural functions of the subject... or in the images that organize at the limit of the *Innenwelt* and the *Umwelt* their relational structuring.[6]

These differences, and their distinct but multiple causality, are worth exploring, but for our purpose here it is sufficient to note that the "body" in psychoanalysis no longer corresponds to the "organism" of biomedical knowledge but must be understood, Freud says, in relation to the image and the word.

This allows us to see more clearly the stakes of Freud's disagreement with Charcot. When it comes to locating the "seat" of disease, and the anatomical "lesion" that accounts for the symptoms of hysteria – as required by the language and conceptual framework of medical discourse – Freud insists on a radical division. Without simply abandoning the obligatory language of medical discourse, Freud nevertheless marks its limit, and the difficulties that have begun to surround it. In the case of periphero-spinal and cerebral paralysis, we have two distinct neurological systems, and the "lesion" in each case would have to be located in one of these two neural systems of "representative fibres." But *where* is the locus of the lesion in hysteria? Does it have a "location" *apart from anatomy*? The hysterical symptom, to be sure, has a precise material form, and one may speak of its "bodily" location. But is the "lesion" that causes this symptom also located in a similar physiological manner? More broadly still, is the body organized as an anatomical space, like other "objects" of natural science, as the discourse of science and the medical language of "lesions" would lead us to suppose? Does the organization of the body follow the "space" of "extended substance" as established by Descartes, with its geometrical coordinates that separate inside and outside (as with a square or a circle); or does the body perhaps present us with another spatiality, in which inside and outside are involuted or unstable (as Freud suggests when he speaks of the "orifice" and the "relation to the object," especially in cases of oral incorporation, where *the first object is recognized* as "good," only to be incorporated by the subject, and *reduced to identification*)?[7] The body, in psychoanalysis, is not automatically integrated as a whole, with a clear inside and outside, as one finds in the case of the animal organism. On the contrary, the "body" must acquire whatever tenuous unity it may possess through the operation of images and language, and the unity that it thereby acquires can no longer be reduced to that of a "thing" in nature, on the model of material extended substance. The problem of the "lesion" in hysteria is precisely this problem of locating the cause of the hysterical symptom, and giving it a "place," but in a manner that violates the expectations of medical discourse. Where, Freud will later ask, is the "unconscious"? Can the unconscious be located anatomically, and photographed by a brain scan, and if not, does this mean that psychoanalysis is only a "pseudoscience" propagated by charismatic charlatans? In what part of the soul does "unconscious memory" reside, particularly if the symptom "remembers" what the memory forgets?[8] How, then, can Freud claim to address the question of the "lesion" in hysterical paralysis, as required by the authorized academic discourse of the day? Does this not amount to abandoning the very field of psychoanalysis, in favor of organic medicine, and the Cartesian understanding of space that underlies it?

Speaking of hysterical symptoms, and the differences that separate them from the symptoms of organic paralysis, Freud writes: "In order to explain this difficult problem, we must certainly consider the nature of the lesion concerned" ("Motor Paralyses" 168). But Freud quickly adds that Charcot's approach allows psychoanalysis to be reabsorbed by the discourse of biomedical knowledge, thereby collapsing the distinction between "hysterical" and "cerebral" paralysis. As Freud observes, "We have several times heard from M. Charcot that [in cases of hysteria] it is a cortical lesion, but one that is purely *dynamic* or *functional*" (168; emphasis mine). This means the lesion in hysteria has *no observable organic location*, but appears (1) only *as an effect* (*dynamis* in the sense of "force": in keeping with an old philosophical dichotomy, the hysterical lesion has "potency" or the "potentiality" to cause effects, without appearing in the body as "actuality"; it has "force" rather than physical "form"), or (2) *as an impairment of some physical function*, such as walking or speaking, but without any visible "seat" apart from this functional disruption. As Freud notes, this "is equivalent to asserting that no appreciable tissue changes will be found *post-mortem*" (168).

Charcot's account of hysteria thus remains attached to a biomedical discourse that does not allow the phenomena of hysteria to appear in their own distinct and challenging difference.

He thereby reabsorbs the emerging domain of psychoanalysis into the more familiar territory of organic medicine, but at the same time leads the language of medicine into a kind of impasse, contradiction, or limit, where it encounters a "lesion" that is *presumed to exist organically*, while *remaining invisible*, operating in a "functional" or "dynamic" way that produces effects, but that cannot be materially located, so that the lesion leaves no trace of itself behind for the anatomist to examine post-mortem.

II the "body" and history of medicine

If one recalls Foucault's analysis of the emergence of modern medical knowledge at the end of the Enlightenment, and in particular the rise of tissue theory described in *The Birth of the Clinic*, the stakes of this disagreement for the history of thought become clearer.[9] For the emergence of tissue theory (so famously outlined by Foucault in connection with the once-scandalous practice of post-mortem anatomy) and the broader emergence of organic medicine as a new biological science in the early nineteenth century allowed medical discourse to abandon the old "speculative" theory of humors, in which "disease" was a kind of abstract entity that entered the body and caused illness, only to disappear again as soon as the body was either cured or killed. This abstract "metaphysical" theory of disease (black bile, phlegm, etc.) was abandoned when anatomy began to describe the specific physical traces that each disease left behind, classifying diseases according to their *organic locations*, along new conceptual lines that no longer conformed to the traditional theory of humors. Tissue theory arose as a result, since each disease could now be shown to have a specific "seat," in some cases progressing from one location to a larger area, in other cases remaining in a particular organ or functional system (such as respiration). This is the anatomical base that Charcot leaves intact, at the same time pointing to its limits, but remaining within its boundaries, without forging an alternative (indeed, he seems to fall back into the "speculative" model of the humors, by claiming that hysteria produces a "functional" or "dynamic" symptom complete with physical effects, but without its actual physical cause ever becoming visible).

Freud takes a different path, and he criticizes his master on just this point, which concerns nothing less than the disciplinary specificity of psychoanalysis – not simply as a *discourse* but also as a *domain of new objects*, which we are tracing here in terms of the hysterical "symptom," and more broadly in terms of the "body" as such. "I am quite sure that many who read M. Charcot's works believe that a dynamic lesion is indeed a lesion, but one of which no trace is found after death" ("Motor Paralyses" 168), Freud writes. "I, on the contrary, assert that the lesion in hysterical paralyses must be completely independent of the anatomy of the nervous system" (169). Maintaining the obligatory language of "lesions," Freud nevertheless insists on a new domain of knowledge for hysteria, a different causality, and a fundamentally different phenomenon when it comes to the hysterical symptom, which is organized not by the laws of anatomy but, as he briefly suggests, in terms of the "image" and the "name."

In the final section of his brief text, Freud attempts to sketch "what the lesion that is the cause of hysterical paralysis might be like" (169). He rejects Charcot's formulations, abandoning the very territory of the "organism" upon which the expectation of a "lesion" is based. He claims instead "that there *can* be a functional alteration without a concomitant organic lesion" (invisible or otherwise), and he argues that we can approach this distinctive type of symptom only by shifting from the organism to another domain: "I only ask permission to move onto psychological ground." Following Pierre Janet, he claims that "what is in question in hysterical paralysis... is *the everyday, popular conception of the organs and of the body in general.* That conception is not founded on a deep knowledge of neuro-anatomy but on our tactile and above all our visual perceptions" (170; emphasis mine). He concludes that if we must maintain the vocabulary of the "lesion" it must be reconceptualized and displaced onto another dimension: "the lesion in hysterical paralysis will therefore be an alteration of the *conception*, the *idea* of the arm, for instance."

Here again, one may note that Freud opens a distinction between the imaginary and symbolic that he does not yet explore in any depth. Based "above all" on "visual perceptions," the hysterical symptom nevertheless also involves an alteration of the "conception" or "idea" of the arm (the logic of the "name"). As has often been pointed out in the history of philosophy, the Greek term *eidos* ("idea") designates both (a) *a concept*, an abstract notion such as "justice," or the "forms" in Plato, and thus a category that only belongs to language, and also (b) *a visual form*, an object one can see or touch. The same double reference reappears in the English "idea," whose root "id" comes from the ambiguous Greek verb *idein* (to "see" or to "know"), remains present in the Latin *videre* ("to see"), and descends to the English words "visible" and "evident." This murky intersection between the image and the name or concept, so crucial to Freud's account of the hysterical symptom, is not yet elaborated with any attention, but it will be a central concern to him in the years that follow, and its philosophical pedigree can be traced back to the Platonic "Forms," which (like "justice") are not empirical objects, but ideas, shining abstractions whose beauty one can only "see" with the "mind's eye." One might therefore wonder what it means to speak of the "form" of the body or "bodily form" (as Lacan does in "The Mirror Stage") as if it were simply reducible to the visible image. Can we really speak of the "body" as an entirely "imaginary" body, as Lacan is so often said to do? Even the text of "The Mirror Stage" suggests otherwise.[10]

The ambiguity of the word *idein*, the visual or linguistic "form," thus reappears in the ambiguity of the hysterical symptom, corresponding as it does to both the "image" and the "concept" of the arm. This problem – which concerns the "logic" of the symptom insofar as it departs from biomedical knowledge – is not addressed by Freud in this text, but it is precisely the peculiar intersection that joins and separates the imaginary and symbolic, a problem Lacan will elaborate at great length. In fact, it is already evident in the traditional vocabulary of the "imago," a vocabulary that Freud takes from broader psychological discourse, and that Lacan uses in "The Mirror Stage" when he speaks not simply of the "imaginary body," as is so often said, but rather of the way in which the imaginary body is taken up and reconfigured by the signifier, in the course of what Freud called "secondary identification." "We have only to understand the mirror stage *as an identification*," Lacan says, an identification which does not *represent* the subject in a visual reduplication (indeed, far from *representing* a pre-existing subject, it *constitutes* the ego and brings it into being, beyond any natural existence) but rather presents us with a transformation, "the *transformation that takes place in the subject* when he assumes an image," a point which "is sufficiently indicated by the use, in analytic theory, of the ancient term *imago*" (2). Lacan is thus quite content to map his terminology (the "imaginary") onto the existing discourse of the new conceptual domain that breaks with Charcot, and he goes on to note that this initial identification "would have to be called the Ideal-I, if we wished to incorporate it into our usual register." But the initial "form" of the ego, what Freud called the "ideal ego," will undergo a second transformation under the heading of symbolic identification, which Freud regarded as the development of the "ego ideal," and it is *this* identification, governed by language according to Lacan, that allows us to explain what Freud calls "sexuality." As Lacan puts it in "The Mirror Stage," the ideal ego "will also be the source of secondary identifications, under which term I would place *the functions of libidinal normalization*" (2). Does this "libidinal normalization" simply amount to a conservative policing of the "proper" object-relation, culminating in the heterosexual normativity that critics have so often claimed to find in Freud and Lacan? A reading of the text suggests that the real problem is quite different from this.

For, in fact, there is a kind of aggression in the human being, a destructiveness that has no counterpart in nature, and cannot be grasped in terms of any Darwinian mechanism of survival, governed by the individual organism, or its genetic afterlife by way of reproduction. The aggressiveness of the ego is thus detached from the biological order of life and death (just as the master–slave relation, in Hegel, entails a risk of death, not a struggle to maintain its organic survival but a risk of that life for the sake of a recognition that exceeds any biological need). This distinctly human aggressiveness that breaks with all natural aggression is the correlate of the ego, and Lacan claims that this is what

> the first analysts tried to describe when they invoked destructive and, indeed, death instincts, in order to explain the evident connection between the narcissistic libido and the alienating function of the I, the aggressivity it releases in any relation to the other. (3)

"Libidinal normalization" thus refers first of all to the process by which this destructive libido finds another path, a way into the field of representation, where it can develop beyond its initial narcissistic investment, finding a way to the object (along paths whose multiplicity Freud repeatedly stressed), and thereby developing "a local habitation and a name," as Shakespeare says in *A Midsummer Night's Dream*, such that desire will have a chance to be pursued beyond the horizon of this initial, paralyzing destruction.[11] This is the task of "secondary identification," which Lacan called identification with signifiers, and which Freud developed under the heading of the "Oedipal Complex." The passage from the ideal ego to the ego ideal (secondary identification) is the elaboration of this trajectory into language which, as it were, detoxifies the ego, understood as contaminated at its very origin, marked by a form of unnatural aggression that may appear as Hegelian "rivalry with the other," but is in fact based on an original division of the subject against itself ("identification with the rival is not self-evident, except at the level of fable, and can only be conceived of if the way is prepared for it by a primary identification that structures the subject as a rival with himself" 22). The problem of "sexuality" is therefore the problem of creating a field in which the libido can escape its initial narcissistic capture, together with the imaginary rivalry that results, in which the "other" is reduced to another self, an "alter ego." This is why, in "The Mirror Stage," Lacan elaborates his reference to the "destructive and, indeed, death instincts" with a remark about "the dynamic opposition between this libido [narcissistic libido] and the sexual libido" (6). Given its initial narcissistic formation, the ego will absorb the libido, which can no longer be regarded as having a natural connection to the proper biological object; and "sexual libido" thereby emerges in opposition to narcissistic libido, and even as a threat. As Lacan says, this

> turns the I into that apparatus for which every instinctual thrust constitutes a danger, even though it should correspond to a natural maturation – the very normalization of this maturation being henceforth dependent, in man, on a cultural mediation as exemplified, in the case of the sexual object, by the Oedipus complex. (5–6)

Lacan returns to this topic in "Aggressivity in Psychoanalysis," in terms that make still clearer the peculiar linkage between the imago and language. The very origin of the ego confronts us with the "notion of aggressivity as a correlative tension of the narcissistic structure in the coming-into-being of the subject" (22). And this initial "tension," together with the rivalry and destruction it entails, must be brought into "dialectical relation with the function of the Oedipus complex," which Freud elaborated in terms of "*secondary identification* by introjection of the *imago* of the parent of the same sex" (22). Lacan will abandon the language of the "imago," and elaborate these secondary identifications in terms of *identification with signifiers.* As a result, the imaginary explanation by which "gender identity" is reduced to images, and explained by means of the purely visual difference between the sexes, will be replaced by another logic. And, of course, this "other logic" of language is already present in Freud, according to Lacan, if only one reads his text – as should already be evident in the "introjection of the *imago* of the parent," where "introjections" includes a process of "psychic representation" that goes well beyond the visual image. As Freud notes in *Totem and Taboo*, the cannibalistic "incorporation" of the dead Father is also a group identification, a belonging that transcends the ego and reconfigures it, such that tribal or clan identity (identification with signifiers) is secured.

> Only the anti-dialectical mentality of a culture which . . . tends to reduce all subjective activity to the being of the ego can justify the astonishment of a Van den Steinen when confronted by a Bororo who says "I'm an ara'" . . . which, on reflexion, is no more surprising than declaring "I'm a doctor" or "I'm a citizen of the French Republic." (23)

Faced with this diversity of identifications, how can readers suppose that Lacan simply reduces identification to a familiar binary opposition? The relations between the imaginary and symbolic are extremely complex, and already apparent in Freud's formulations on the peculiar, non-biological logic by which the hysterical symptom is formed. For our purposes here, however, with respect to the problem of the "body," it is enough to stress the crucial role of "libido" in connection with the problem of aggressivity. For "libidinal normalization," whatever it may mean in psychoanalysis, clearly concerns a modification of the unnatural destruction that arises with the genesis of the ego. This is what Lacan calls the "pacifying function of the ego ideal": "What concerns us here is . . . the pacifying function of the ego ideal, the connection between libidinal normativity and a cultural normativity bound up from the dawn of history with the *imago* of the father" (22).

In his text on organic and hysterical paralysis, Freud resolves these ambiguities largely in favor of the signifier. What distinguishes the hysterical symptom from its organic counterpart is the former's link to the "conception" of the arm – its symbolic as distinct from its biological dimension. This requires a displacement of the classical medical understanding of the "lesion." "The lesion," Freud writes, "would therefore be the *abolition of the associative accessibility of the conception of the arm.* The arm behaves as though it did not exist for the play of associations" ("Motor Paralyses" 170). This is why one finds no organic damage, once the symptom has been resolved by the talking cure: the arm can be "inaccessible" to the play of representations, and can thus lead to a functional alteration, but without any underlying organic damage ("without being destroyed and without its material substratum (the nervous tissue of the corresponding region of the cortex) being damaged" (170). In short, Freud concludes that, if the "lesion" in hysteria has no anatomical location in the sense required by organic medicine, while nevertheless allowing the symptom to materialize in the body, this is because the "lesion" concerns a rupture in the chain of psychic representations.

We have passed from the domain of organic medicine to the domain of psychoanalysis. From this point on, we must distinguish between the "organism" (which belongs to medical discourse and remains undamaged in hysteria) and the "body" (where the symptom is now located). And as Freud will go on to argue, the "body" is the correlate of the ego, rather than being a pre-existing natural entity that biological knowledge might describe. As Freud already says in this early text, "The conception of the arm cannot enter into association with the other ideas constituting the ego, of which the subject's body forms an important part" (170).

These remarks will, of course, be famously developed by Lacan in his account of the "imaginary body," in its radical departure from the biological organism, but the same correlation between the body and the ego is fully evident in Freud. Consider, to take only one example, Freud's texts on anxiety, where he argues that "the ego is the actual seat of anxiety."[12] Here again, Freud relies on the traditional language of organic medicine, which requires a "seat" (or organic location) for every disease, but he simultaneously dislocates that received language by situating it in a new domain: "Anxiety is an affective state and as such can, of course, only be *felt by the ego*" (140; emphasis mine). What does it mean to speak of a "feeling" in the ego? It cannot simply be a question of *sensation*, understood as a biological capacity to "feel" (touch, hearing, sight, etc., conceived as biological senses). It is certainly not a concept, and it cannot be located at the level of the signifier. Indeed, in his article on "The Unconscious," Freud famously notes that clinical practice had long been preoccupied by language, and the labor of "free association," but that clinical observation now requires that the analyst "follow up separately" what happens at the level of unconscious thought, and what happens at the level of affect. Whether one calls this "feeling in the ego" an "affective state," a "mood," a "state of mind," a "disposition" or "temperament" of the subject – all terms that were developed by Heidegger and others, and that have a strong genealogical foundation in Kant and the "feeling in the mind" that characterizes esthetic pleasure, as a distinctively "disinterested" pleasure that is distinct from organic sensation or "appetite," and at the same time distinct from and conceptual or linguistic determination – it is clear that the "affect" of anxiety is one that cannot be grasped in terms of the biological organism.[13] Anxiety is now an attunement or disposition of the subject, and is "felt" by the ego rather than by any sensory apparatus. This places anxiety in a domain that cannot be reduced to biology. Whatever one may wish to say about the neurological pathways that regulate tension and discharge in the organism, and however prominent may be the tremendous effort of pharmaceutical corporations, the medical industry, and other institutions of knowledge to regulate these pathways by chemical means, Freud insists on a different path. As he says in his *Introductory Lectures on Psychoanalysis* of 1917, in the chapter titled "Anxiety":

> You will certainly expect psychoanalysis to approach this subject too in a quite different way from academic medicine. Interest there seems mainly to be centered on tracing the anatomical paths along which the state of anxiety is brought about. We are told that the medulla oblongata is stimulated, and the patient learns that he is suffering from a neurosis of the vagus nerve. The medulla oblongata is a very serious and lovely object. I remember quite clearly how much time and trouble I devoted to its study many years ago. Today, however, I must remark that I know of nothing that could be of less interest to me for the psychological understanding of anxiety than a knowledge of the paths of the nerves along which its excitations pass.[14]

Thus, if the ego is the "seat" of anxiety, that affect is also manifested in the body, insofar as the "body" is the correlate of the ego.[15]

In short, Freud's early text on the hysterical symptom deserves to be read as a decisive moment in the emergence of psychoanalysis as such. It testifies to the emergence of *a new type of thinking* about the body, and more than this, to a new conception of *the body itself*, as distinct from the "organism" that is addressed by biomedical theories of disease. In other words, it is not a matter of deciding whether psychoanalysis has a more correct or proper account of the body than medical discourse but rather of recognizing a new division in the history of thought, a division between two kinds of symptoms, one of which has a biological cause, while the other has a real physiological and material existence, but with a cause whose logic remains to be determined. From this perspective, it is not difficult to see that when contemporary psychoanalysis tries to justify itself against the proliferating attacks of those who insist that there is a simple choice to be made about psychoanalysis, and that psychoanalysis must either prove that it is a "genuine science," or confess that it is only a form of mysticism grounded in the charisma of its practitioners, we can see that the very emergence of a new and distinct domain of knowledge is being eliminated – interpreted in terms of a traditional alternative (science vs. mysticism) that leaves no room for the specificity of psychoanalysis as a discipline.

It should therefore be clear that psychoanalysis still holds a powerful and disruptive potentiality with regard to questions of sexuality and gender, vis-à-vis other more well-known discourses, particularly biological models and models grounded in what is called "the historical construction of sexuality." Psychoanalysis by no means discredits those more familiar models, which rest on long-standing disciplinary foundations of natural science and historical analysis. But as Foucault himself clearly recognized, in "Preface to Transgression," his famous discussion of Georges Bataille, psychoanalysis is a newly emerging discourse, a form of thought and practice that is irreducible to its predecessors.[16] As we argued at the beginning of this essay, psychoanalysis remains clearly distinct from the forms of thought that prevailed in Freud's day, but at the same time it runs the perpetual risk of disappearing when it is absorbed by more familiar modes of knowledge – and this risk is visible even in Freud himself.

III the truth of psychoanalysis

If we insist on the distinction between the "organism" and the "body," then, it is to leave room for the articulation of two distinct approaches, two approaches which are not rivals but which rather direct themselves to two different objects. Put differently, we might say that the "truth" of psychoanalysis will always disappear when it is displaced onto another domain – whether it be the domain of natural science, or that of cultural studies and historical construction. And since the reception of psychoanalysis has largely been characterized by just such a displacement or translation, we might say that psychoanalysis always risks disappearing, and that its intellectual life may even depend on this capacity to survive translation. Psychoanalysis, and the "truth" of psychoanalysis, is quite distinctive in this respect, compared to other forms of knowledge. Scientific discourse, and literary discourse (at least in its traditional "formalist" form), have very clear procedures and boundaries. Obviously one can show that science develops in a radical dependence on culture and language, but from the standpoint of the scientist, the procedures of scientific truth are regulated by internal scientific standards; similarly, literary works can certainly be shown to be deeply interwoven with cultural, political, religious and other values, but from the standpoint of traditional formalist criticism the literary object itself should be understood from within the universe of literature and not read as if it were a historical or ideological document. But in relation to scientific and literary discourse psychoanalysis is, in a certain way, homeless or vagrant, easily displaced and absorbed by other forms of thought. This does not mean that psychoanalysis has no disciplinary specificity, but that its actual existence as a discourse (and even as a practice) is strangely porous, or hospitable, or vulnerable to displacement. Its institutional life is curiously close to that of religion, its intellectual kinship is closely linked to humanistic disciplines, and its ideal self-conception is scientific, and its practice involves a singular experience that is comparable to almost nothing else. One has only to look at the itinerary of Freud's own work – his path through anthropology, medicine, religion, literature, etc. – to recognize what we are calling the nomadic or vagrant character of his thinking. Freud's thought does not migrate through foreign territory simply to demonstrate an "extrinsic" relation to other disciplines – as philosophy may do when it tries to produce a "deconstructive reading" of a literary text, or as literature may do when it exports its methods to anthropology, allowing the cultural world to be read as a "text." Instead of starting with his procedures and concepts well established, and taking them out to conquer other domains, Freud actually turns to anthropology and literature in order to discover something that will allow psychoanalysis to progress. In this sense, psychoanalysis depends radically on its "outside" in a very distinctive way, as if it had no clear boundaries for its own disciplinary development. At the same time, however, something of psychoanalysis survives within this very movement of migration or hospitality. Far from being lost or simply displaced by this migratory tendency, psychoanalysis finds itself and progresses in Freud's work precisely as a result of this conceptual mobility. Far from losing itself, it may even be that psychoanalysis "itself" could not survive without this curious reliance on neighboring forms of thought, and that, unlike other discourses which remain within their own well-established boundaries, psychoanalysis is a form of thought that depends on this relation to its neighbors, not as an occasional wayward movement beyond its proper boundaries but as the very condition of its own articulation. Freud often suggested as much when he developed his theoretical work by taking psychoanalysis outside its proper domain, into the territory of anthropology, religion, literature (Greek tragedy), and even speculative evolutionary theory.

How, then, should we think about "the truth of psychoanalysis," or truth "in" psychoanalysis, if psychoanalysis indeed has a clear boundary, an "inside" that would separate its truth from that of its disciplinary neighbors? And how can psychoanalysis maintain the distinctive territory opened up by Freud's work, in the face of this strange hospitality to other forms of thought? For on the one hand, a major tendency – within the largely medicalized horizon of institutionalized psychoanalysis – has been to seek legitimacy by bringing psychoanalysis into line with the traditional expectations of scientific discourse. On the other, in the arena of cultural studies, where psychoanalysis has had the most success in obtaining recognition and interest, the discourse of psychoanalysis has been integrated within a broader network of theories about gender, social construction, ideology, interpellation, identification, desire, and forms of "representation" and "identity" that have relatively little connection with the terminology and conceptual framework of psychoanalysis. This blurring of conceptual boundaries is often extremely productive, and should by no means be simply denounced, as if the proper task of the psychoanalytic critic were to police the boundaries of its discourse, with the aim of expelling all foreigners or intruders because they speak a different tongue. On the contrary: it might well be that psychoanalysis intrinsically invites this very blurring of intellectual boundaries, and that psychoanalysis can be characterized as more hospitable and more open to intrusion than other discourses (science, religion, and even traditional (formalist) literary criticism), where proper boundaries and procedures are more securely established.

Nevertheless, it is important for readers of Freud and Lacan to remain attentive to those moments when the distinctive features of psychoanalysis are forgotten and erased, or when psychoanalytic truth is reabsorbed by more familiar forms of knowledge. As Foucault said on many occasions, history always tends to be written by the conqueror, and constructed as a narrative of progress leading to the present. It is much easier for us to speak of psychoanalysis in terms of the discourses that are most strongly established in our culture today, and to view it either (1) as a form of "social construction" which finally denounced the old truths of biological essentialism, so that psychoanalysis can be absorbed by cultural studies and other discourses on "symbolic" identity, but only at the cost of leaving its peculiar vocabulary behind (the unconscious, libido, the drive, etc.), or (2) as a failed science that maintains its flagging prestige only in faddish academic circles that, being mainly literary, do not maintain any genuinely scientific requirements for truth – fields that tend to be captivated by intellectual fashion and the charisma of those "creative" figures who, like Freud and Lacan, but also Derrida and Deleuze, become the momentary cultish leaders of these disciplines. On this account, and whether it succeeds or fails, psychoanalysis has already disappeared, having been absorbed by rival discourses that celebrate or denounce it, but no longer maintain its unique discursive field. As Foucault often said, every genuine rupture in knowledge, every genuine creation, will always risk being reabsorbed by more familiar neighboring discourses, and this is why he claims in 1963 that, despite Freud's writings, the psychoanalytic discovery of sexuality, and the language in which sexuality can begin to emerge and articulate itself, remains almost entirely in the future.

IV sexuality and the "law"

This is perhaps especially important with respect to the question of ethics. According to Foucault, the psychoanalytic experience of sexuality, and the account of "transgression" which it entails, "must be detached from its questionable association to ethics if we want to understand it and to begin thinking from it and in the space it denotes; it must be liberated from the scandalous or subversive."[17] "Transgression," Foucault writes,

> does not seek to oppose one thing to another, nor does it achieve its purpose through mockery or by upsetting the solidity of foundations . . . Transgression is neither violence in a divided world (in an ethical world) nor a victory over limits (in a dialectical or revolutionary world). (35)

In these words, the historical position of psychoanalysis is conspicuously marked by Foucault. The location of this "singular experience" of transgression has to be understood as a rupture with the forms of thought that organized religious truth ("violence in a divided world" where good and evil, sin and obedience, or the sacred and profane are the functional categories), as well as a rupture with the models of transgression developed by dialectical thought (revolution, reversal, and dialectical contradiction). The experience of sexuality after Freud is the emergence of an experience that cannot be reduced to these prevailing and more familiar models of thought.

After the death of God, transgression can no longer be understood as a crossing of the limit, the limit of nature or the limit of the Law. Since sexuality has no norm – biological or cultural – by which it can be measured, and thus no norm against which it can transgress, it necessarily involves a new experience of transgression, which can no longer be characterized in the traditional way, as a crossing of the limit or as a "breaking" of the law; instead, transgression – the transgression that sexuality is, not in its extreme form, but as such, and in its very nature – must be understood as a movement, or an experience, that no longer belongs to the traditional categories of "inside" and "outside," whereby one might either "follow" or "transgress" some established limit. "Transgression," Foucault writes, "is not related to the limit as black to white, the prohibited to the lawful, the outside to the inside" (35). As an experience that has no "proper" form, sexuality introduces a form of transgression that no longer belongs to the logic of inside and outside, lawful and prohibited. In place of dialectical contradiction, which requires a subversion or reversal (*Aufhebung*) of established knowledge, sexuality produces an experience in which the law and transgression intrinsically belong to one another, no longer simply "oppose" each other, but entail a new relationship in which the two belong together, without contradicting each other.

How can we think this relationship, which is not exactly that of contradiction, but in which the limit and transgression belong together, rather than functioning as an opposition? How can we "understand it and begin thinking from it, and in the space that it denotes" (35)? Given Foucault's observation that sexuality "must be detached from its questionable association with ethics," one might wonder about the distance that separates this conception from more familiar models of critical practice in which "transgressive" or "deviant" or "non-normative" sexuality are enlisted as a form of cultural critique. Such models, appealing to alternative or marginal forms of sexuality as a form of contestation aimed at displacing cultural norms, are bound up with the very ethical practice that Foucault questions here, when he claims that sexuality, and the peculiar transgression it entails, "must be detached from its questionable association to ethics if we want to understand it and to begin thinking from it and in the space it denotes; it must be liberated from the scandalous or subversive" (35). This is not to say, by any means, that there is no place for forms of cultural critique that engage with questions of sexuality, gender normativity, and the subversion of established cultural practices. It is only to mark the difference, which I take Foucault to be underscoring here, between the psychoanalytic account of sexuality after the death of God, and forms of cultural critique that enlist alternative sexualities as "contestatory" forms, using them to criticize prevailing norms, in the name of ethical discourse – such forms of analysis remain bound to more dialectical models of "critique," in which and inside and outside, or law and transgression, remain opposites that stand apart from one another. One familiar version of this model argues that Lacan seeks to maintain a particular "normative" form of sexuality, and that this norm can be contested "from outside" as it were, by an alternative possibility. If Foucault is correct, this critical and oppositional discourse is historically "outside" the discourse of psychoanalysis, not only because it continues to think transgression in a more traditional way but also because it remains bound to forms of ethical practice that the contemporary experience of sexuality should, according to Foucault, disrupt. Sexuality, Foucault says, "must be liberated from the scandalous or subversive," for the experience of transgression that it entails "does not seek to oppose one thing to another, not does it achieve its purpose through mockery or by upsetting the solidity of foundations" (35).

A similar gesture is made in the famous opening of *The History of Sexuality*. It looks as if the Victorians repressed sexuality, while modern thought, following Freud, Bataille and others, has finally liberated it. This, at least, is the well-established history that Foucault sets out to question, by constructing a counter-history. But the main emphasis of his analysis is not on the past; his aim is not primarily to provide a more accurate narrative about our ancestors, but rather to provide a genealogy of the *present*, that is, to counter the current prevailing forms of discourse in which sexuality is enlisted as a liberatory practice, and attached to ethical values that Foucault finds inadequate. It is not the oppressiveness of Victorian life that interests Foucault, or even the secretly scandalous reality of that time, but rather *our own story*, the narratives that we construct about our ancestors. For these narratives organize our understanding of ourselves, and give us the privilege of a kind of ethical superiority, thereby holding us hostage to a truth that is inadequate. As usual, Foucault is not simply asking about the past but also asking "Who is speaking" in the received narrative. For the usual story of repression among the Victorians, followed by liberation in our own day, bears a remarkable similarity to the grand narratives in which modern psychiatry is said to overcome the barbaric and ignorant practices of the past. In short, the narrative of repression among the Victorians has its reasons, and "is easily analysed," for "the sexual cause – the demand for sexual freedom... becomes legitimately associated with the honor of a political cause." There is a reason, Foucault writes,

> that makes it so gratifying for us to define the relationship between sex and power in terms of repression: something that one might call the speaker's benefit. If sex is repressed... then the mere fact that one is speaking about it has the appearance of a deliberate transgression... Our tone of voice shows that we know we are being subversive, and we ardently conjure away the present and appeal to the future, whose day will be hastened by the contribution we believe we are making. Something that smacks of revolt, of promised freedom, of the coming age of a different law, slips easily into this discourse on sexual oppression. Some of the ancient functions of prophecy are reactivated therein.[18]

It is clear, then, that Foucault's history is aimed primarily at the present, and intended to displace the standard practice in which sexuality is associated with transgression in its traditional form, and with the ethical framework of the past, in which liberation is attached to subversion, the lifting of a repression or the crossing of a normative limit. Sexuality today, after Freud, can no longer be understood if it is contained within that traditional ethical framework. A different form of thought and practice must be allowed to emerge, and we must detach our passions – our conceptual and libidinal investments – from those old ideals. Foucault himself, of course, is practicing a form of contestation in this text, as in so many others, and one might call his work a form of "critique." But I would argue that this practice has more to do with the displacement of received knowledge, and what I would call the effort to exhaust familiar paradigms, than it has to do with traditional forms of critique, dialectical or otherwise. In fact, Foucault's way of reworking received narratives, retelling the history that has made us who we are today, and reworking those stories in such a way that their meaning is gradually exhausted – so that they no longer bear the truth for us, and they no longer function, and no longer support our desire – all this is closer to the work of psychoanalysis than it is to philosophical forms of critique that remain indebted to the Enlightenment.

It seems to me that this new relationship between law and transgression, described by Foucault, was developed powerfully by Lacan in his account of sexual difference, where the law cannot be understood apart from its own internal malfunction. In Lacan's account, it is not a question of a law that one can follow or transgress. It is not a question of existing as a "normal" subject, with a "proper" object choice, a heterosexual normativity, etc. or as a "non-normative" subject whose forms of life would violate the norm. What Lacan showed was rather that the "normal" subject is intrinsically faced with an internal contradiction (or "division," if one wishes to avoid traditional logical opposition at all cost). This contradiction is expressed in many ways. It appears, for example, in the very title of Freud's work *Civilization and its Discontents*, which indicates that the very submission to "civilization" (not simply the moral codes of a culture, for Lacan, but the very fact of language) brings with it a dimension of ineradicable "discontent." Here, accepting the law also means that one internalizes or experiences a measure of discontent that contradicts the law itself.[19] As Freud notes in *Totem and Taboo*, it is no longer the "outlaw" father who represents a transgression of the law (since he has been killed, according to the famous story); rather, it is the sons themselves who, from the very fact that they agree to abide by the law, and to constitute a sort of democratic brotherhood in which none of them will seek to occupy the "transgressive" position of the primal father, become paradoxically the most guilty, the most tormented by superego guilt. On this account, the law is no longer transgressed by the outlaw figure of the father who refuses to sign the social contract (a model familiar to us from Hobbes, Rousseau, and other social contract theorists); rather, the law is transgressed within itself, not by some outsider, but in the experience of those who follow the law most faithfully, the sons, the very subjects of the symbolic order.[20] In Lacan's account, of course, this characterization of the law as intrinsically entailing its own contradiction or self-disruption is further complicated by an alternative feminine position in which there is no outside to the law, no primal father, and at the same time no complete inscription within the law. The law is thus equally incomplete or unstable on the feminine side, but in a manner that is distinct from what one finds under the heading of masculinity.[21] I will not develop these points here, but only wish to stress that in either case we are faced with a conception of the law that intrinsically entails its own violation or incompleteness – not because the law has failed to operate properly or completely but, on the contrary, because the very operation of the law necessarily entails this internal disruption.

It would be foolish to pretend that Lacan, or indeed Foucault, is the only thinker who has approached this problem of a law that entails its own transgression. Derrida was repeatedly and even obsessively concerned with this problem – whether it appears as the paradoxical relation between writing and speech, in which a familiar dichotomy turns out to involve a more complex and paradoxical relation, such that speech is always already tangled up with its own opposite, so that the limit between inside and outside is destabilized, or whether it appears in the later work, as a concern about the aporetic structure of the law, and the curious way in which every law contains its own impossibility, its own transgression or incompleteness.[22] Derrida's work on justice is one of the best-known examples of this problem. In "Force of Law," his text on Walter Benjamin, "justice" is at once required, called-for, and necessary, and at the same time impossible, since no written code, no instituted law, will ever bring about justice. The law as such, in its finite, historical instantiation, according to Derrida, always contains some measure of violence and exclusion, which means that justice must always be demanded, as an infinite task that can never be accomplished. Justice is therefore always necessary and always impossible, in contradiction of traditional philosophical logic, which would oppose these categories to one another.[23] Like the ethical relation in Levinas, to which Derrida's formulations are deeply indebted, justice can never be established or accomplished, and remains an infinite task. It is therefore no longer a matter of regarding the moral law as a principle or commandment that one might follow or disobey – a moral law that one can either obey or transgress. Ethics is no longer a matter of a norm that might be transgressed in the name of the scandalous or the subversive. It is rather a relation to this law, which contains its own impossibility. In fact, as his work progressed, Derrida came to characterize deconstruction itself as both necessary and impossible – a position that one might call melancholic, as I have argued elsewhere (21). Nor do I wish to pretend that Foucault, Lacan and Derrida all take precisely the same position with respect to the problem of law and transgression – a claim that would clearly be untenable. I only wish to suggest that the same historical rupture with dialectical models of contradiction that we find in deconstruction is sketched by Foucault with respect to "sexuality," and developed at great length in Lacan's reading of Freud.

That is not to say that dialectical or other oppositional conceptions of critique have no place or function in contemporary life. Nor is it to say that biological models of thought have no legitimacy or place in contemporary thought. It would be foolish to pretend that psychoanalytic discourse has discovered a truth that puts all other truth to rest. And Freud certainly does not pretend that the discovery of the peculiar features of the hysterical symptom should somehow put medical discourse to rest as a kind of anachronism. On the contrary. It is simply a question of respecting the specificity of the various discourses that operate around us today, addressing themselves to different objects, and demanding our attention at different times in different ways.

notes

1 Sigmund Freud, "Some Points for a Comparative Study of Organic and Hysterical Motor Paralyses" in *The Standard Edition of the Complete Psychological Works of Sigmund Freud* (henceforth SE), eds. and trans. James Strachey *et al.* (London: Hogarth, 1953) 1: 157–72. Subsequent page numbers are given in the text.

2 For further discussion see Charles Shepherdson, "The Role of Gender and the Imperative of Sex" in *Supposing the Subject*, ed. Joan Copjec (London: Verso, 1994) 158–84.

3 See idem, "The Epoch of the Body: Need and Demand in Kojève and Lacan" in *Perspectives on Embodiment: The Intersections of Nature and Culture*, eds. Gail Weiss and Honi Haber (New York: Routledge, 1999) 183–211.

4 See idem, *Vital Signs: Nature, Culture, Psychoanalysis* (New York: Routledge, 2000).

5 Jacques Lacan, "The Function and Field of Speech and Language in Psychoanalysis" in *Ecrits: A Selection*, trans. Alan Sheridan (New York: Norton, 1977) 81.

6 Ibid. 69.

7 See Freud's remarkable text "Negation" in SE 19: 234–39.

8 In his early seminars, Lacan follows these esoteric questions about the "location" of the unconscious, following Freud's own use of optical models in some detail, eventually revealing how Freud abandons these models in the face of the problem of representation in language. For a discussion of this issue with respect to the "anatomical location" of memory, see Charles Shepherdson, "Vital Signs: The Place of Memory in Psychoanalysis," *Research in Phenomenology* 23 (1993): 22–72.

9 Michel Foucault, *The Birth of the Clinic: An Archaeology of Medical Perception*, trans. Alan Sheridan (New York: Random, 1973). Originally published as *Naissance de la Clinique. Une archéologie du regard médical* (Paris: PUF, 1963).

10 In "The Mirror Stage" Lacan argues that the first formation of the ego is restructured with the arrival of language, such that even "libidinal normalization" is affected. "This jubilant assumption of his specular image," Lacan writes, shows us the ego only in its initial form, prior to the dialectic of intersubjective recognition, a different structure which will in turn be modified by language: "the *I* is precipitated in its primordial form, before it is objectified in the dialectic of identification with the other, and before language restores to it, in the universal, its function as subject." Jacques Lacan, "The Mirror Stage as Formative of the Function of the I as Revealed in Psychoanalytic Experience" in *Ecrits: A Selection*, trans. Alan Sheridan (New York: Norton, 1977) 2. Subsequent references to *Ecrits* are in the text.

11 William Shakespeare, *A Midsummer Night's Dream* V.i. The adjacent lines are worth recalling, since they include the poetic gift of creation, and the poet's use of imaginative "forms," which must first be seen by the poet's peculiar "rolling" and "frenzied" eye, and then named or reconfigured by the pen:

> The poet's eye, in a fine frenzy rolling,
> Doth glance from heaven to earth, from earth to heaven;
> And as imagination bodies forth

The forms of things unknown, the poet's pen
Turns them to shapes and gives to airy nothing
A local habitation and a name.

12 Sigmund Freud, *Inhibitions, Symptoms and Anxiety* in SE 20: 140.

13 See Immanuel Kant, *Critique of Judgment*, trans. James Creed Meredith (Oxford: Oxford UP, 1952); and Martin Heidegger, *Being and Time*, trans. James MacQuarrie and Edward Robinson (New York: Harper, 1952).

14 Sigmund Freud, *Introductory Lectures on Psycho-analysis* (Part III) in SE 16: 393.

15 For further remarks see Charles Shepherdson, Foreword to Roberto Harari, *Lacan's Seminar on "Anxiety": An Introduction*, trans. Jane C. Lamb-Ruiz (New York: Other, 2001) ix–lxii.

16 Michel Foucault, "A Preface to Transgression" in *Language, Counter-Memory, Practice: Selected Essays and Interviews by Michel Foucault*, ed. with an introduction by Donald F. Bouchard (Ithaca, NY: Cornell UP, 1977) 29–52.

17 Ibid. 35. Subsequent page numbers are given in the text.

18 Michel Foucault, *The History of Sexuality: An Introduction*, trans. Robert Hurley (New York: Vintage, 1976) 6–7.

19 Sigmund Freud, *Civilization and its Discontents* in SE 21: 64–148.

20 Idem, *Totem and Taboo* in SE 13: 1–255.

21 For further discussion see Charles Shepherdson, "Derrida and Lacan: An Impossible Friendship?," spec. issue on "Psychoanalysis and Social Change," *Differences: A Journal of Feminist Cultural Studies* 20.1 (2009): 40–86.

22 Ibid.

23 See Jacques Derrida, "Force of Law: 'The Mystical Foundations of Authority'" in *Deconstruction and the Possibility of Justice*, eds. Drucilla Cornell, Michael Rosenfeld, and David Carlson (New York: Routledge, 1992) 3–67.

hoon song

TWO IS INFINITE, GENDER IS POST-SOCIAL IN PAPUA NEW GUINEA

One is too few but two are too many.
Marilyn Strathern, Partial Connections *36*

Jacques Lacan's thesis of the "impossibility of sexual relation" has all the features of what anthropologists could have called the "impossibility of species relationship."[1] When examined through the aid of the latter, broadly conceived, I will argue that Lacan's thesis seems already "post-social," and can be read side-by-side with Gilles Deleuze's theses on the social.

The impossibility of the species relationship is reportedly widely held throughout indigenous Amazonia and elsewhere.[2] Eduardo Viveiros de Castro illustrates this as follows:

> what jaguars see as "manioc beer" (the proper drink of people, jaguar-type or otherwise), humans see as "blood." Where we see a muddy salt-lick on a river bank, tapirs see their big ceremonial house . . .[3]

Such a difference in "perspective," he continues, does not derive from difference in the "soul," as the Amazonians stress. That is what people of the West tend to think and he calls this bias *multiculturalist uninaturalism* (1). Our multiculturalist faith in the diversity of representational volition is purchased at the price of an assumed unity of nature. Among Amazonians, it is the other way round. The soul for them is the common original ground of being, regardless of species. Difference is instead located "in the bodily difference between species." For the "body and its affections . . . is the site and instrument of ontological differentiation and referential disjunction" (4). In sum, we tend to think of diversity in terms of "partial representations" of a universally shared objective reality; Amerindians assert a phenomenological unity that is "purely pronominal in kind applied to a real radical diversity" (4). He calls the latter *unicultural multinaturalism.*

One wonders, then, how the question of sexual relation should proceed among Amerindians, that is, under the condition of unicultural multinaturalism. The question sought thereby is not about the "intersection" between species relation and sexual relation, a question one might begin to address by amassing case studies in Amerindian societies. The real weight of the problem rather lies with the "impossibility" as such. What can we learn from the Amazonian "impossibility," if it is indeed a radical thesis that corrects our multiculturalist assumptions? And how can we enlist that lesson in the service of the question of sexual relation (in the multiculturalist world)?

The impossibility in species relation is enlisted here because assumptions that we might call cultural often get in the way. This includes the performative theory of gender *à la* early Judith Butler.[4] Performativism predicates a possible counteraction upon expected social norms, whose negation is believed to result in performative constructs and creativity. So-called "felicity conditions" is one such norm, which itself is gender neutral.[5]

The Amazonian trans-specific "impossibility" cannot be more different. It is not about negating expected norms. It is about affirming relationality where none is expected. It is an *affirmation of impossibility* as such. What achieves that feat, Viveiros de Castro calls the body. The body for him is a "vehicle" (my term) that "pronominally" traverses inter-species difference, without a mediation by, or negation of, a species-neutral "third."

The theoretical intent of the present essay must be apparent in the way "impossibility" (a Lacanian signature) and "affirmation" (that of Deleuze) are uttered in the same breath. I believe such anti-multiculturalist anti-uninaturalism is also the concern shared by the two philosophies of multiplicity(-cum-sexual-difference) that we are called upon to engage here – psychoanalysis and vitalism. Prior endeavors of similar comparison had varying successes in keeping the habit of multiculturalist uninaturalism at bay. More often than not, the pitting of psychoanalysis-vs.-vitalism treated the two as though they were competing commentaries on the same reality, the same "nature." Evidence of this can be found in the prevalence of the either/or conclusion in such comparisons, as dramatized in the title of one study: "You Can't Have it Both Ways: Deleuze or Lacan."[6] It is as though the either/or negation, strictly eschewed in the theories of multiplicity, nevertheless remains unexamined in the *practice* of the same theories. To think against sexual or gender neutrality is to set aside such either/or logic. This includes "psychoanalysis-or-vitalism."

Situating these theories in Amazonias is the experiment at hand. For the bottom line is that reality is "impossible," wherever it is. In Amazonias, this trans-historical lesson stays explicit. That is to say that, there, the multiculturalist-uninaturalist temptation effectively stays deterred. In this essay, I seek that multinatural deterrence not only from Amazonia, however, but more from Papua New Guinea, for the reasons imparted below. Given the multinatural deterrence, then, what might the theories of Lacan and Deleuze look like there? What would Melanesians say about these if they cared to hear the two out? Deterred from the either/or, as I will entertain, we can envision a more nuanced engagement between psychoanalysis and vitalism there. In turn, this engagement offers more productive ways to appreciate Melanesian specificities.

My intent is not to promote Melanesians as would-be theorists, as approximations to the supposed real theories. The thesis of multinaturalism undermines our proverbial notion of theory in the first place. "Approximation" can be said, as a proverbial discourse might, only when presupposing uni-nature. Interposing the "impossible," the thesis also disrupts an easy supposition of filial resemblance – between original and copy, between the theorizing West and the theorized non-West.[7] Under the jurisdiction of multinaturalism, we can say no more that the non-West only "intimates" while the West properly theorizes; that the former only objectifies in "life-worlds" while the latter signifies discursively.[8] In fact, the reverse might be true: I see Lacan's and Deleuze's efforts as aimed at re-animating precisely those insights that remained explicit in Amazonias – insights that were neutralized and mediated under capitalist modernity.[9] That insight, for Viveiros de Castro, is the centrality of the body. The body, for him, is that which enables relationality where there is no relation, where relation is "impossible." Such is relationality that is neither filiationally warranted nor conventionally prescribed. Of such a relation, we can only say "impossible and yet there it is," to borrow Georges Bataille's descriptor for miracle.[10]

The "miracle" of sociality has been the very stuff of anthropology's preoccupations. The socius is "culturally" constructed but precisely as that which is given, as though "naturally."[11] I understand Viveiros de Castro's idea of multinaturalism as a renewed appreciation of culture's "auto-affective" giving of itself the appearance of "nature." His species-relating "impossibility" captures such a self-referential fissuring: "culture's" enfolding of "nature" into itself. Read through Lacan and Deleuze, I will argue, a multinaturalist anthropology cannot but be an anthropology of sexual relation – that is, of the self-fissuring into two of the social.

comparativism, plural and infinite

Papua New Guinea is the site of one of the most conscientious anti-multiculturalist anthropologies ever produced: Marilyn Strathern's *The Gender of the Gift* (1988).[12] My focus here will be on the historical time of Strathern's work, the late 1980s of post-second-wave feminism. Her study is a response to how anthropological comparativism is deployed by second-wave feminist scholars. The reigning mode of scholarship at that time was to deploy indigenous ethnographic specificities to demonstrate the "inapplicability of this or that particular Western concept" (GG 12). What seems to trouble Strathern about the scholarship is its assumption of comparability. An assumed comparability, in turn, presupposes two things. First, a pluralism that is contingently "open"; ethnography-theory comparison can go on indefinitely insofar as one has provisions of ethnos. Second, such an omni-ethnic comparability is in turn enabled by tacitly assumed (uninatural) constants, such as the in-dividual person whose societal "roles" include gender (GG chapters 1 and 2).

In place of plurality, Strathern proposes what we might call infinity. At least that is the term she would have used had she been able to incorporate another Melanesian ethnography which was simultaneously in production at the time of her own: Jadran Mimica's *Intimations of Infinity: The Mythopoeia of the Iqwaye Counting System and Number* (1988). She duly laments as much.[13] What would be the relation between infinity and twoness (of gender)? Why, if I may continue my conjecture, would she have likely thought that infinity, replacing plurality, could free our understanding of gender relation from the grip of multiculturalism?

Strathern would have found it inspiring that the Iqwaye notion of infinity is "closed." Their counting begins with 1 and ends with 1; 1 is both the smallest and the largest number. This is a radical departure from the multiculturalist pluralism's contingent "openness." This openness is consistent with number theory's conception of infinity as a finite-to-infinite *divergence.*[14] Iqwaye's number, in contrast, *diverges* and *converges* at once, so to speak. Accordingly, both open and closed, its infinity has two "sides": (1) it can be approached from "inside" by counting *ad infinitum*; (2) it can be contemplated from "outside" as if counting had already been completed.[15] As such, the Iqwaye infinity has two dimensions: the *yet-to-be-completed* dimension we might call "culture" and the *already-completed* dimension we might call "nature." The Iqwaye infinity inscribes the self-fissuring-into-two of the socius. Strathern would have embraced such an idea of infinity in order to displace both the "cultural"-performativist idea of gender as well as that of the "nature"-determinist idea of sex.

relation, which happens to be sexual

The Gender of the Gift is an effort to render the elements of ethnos uncountable or "deterritorialized." The goal of that book may as well be summarized as "transcendental empiricism."[16] Made transcendental, the twoness of gender difference, for Strathern, is no longer a comparable empirical given, whose unity-in-difference is in turn presumed to be sheltered by a gender-neutral "third," be it culture or society. To that end, she needed a concept of twoness that is more than that which conveys the experience of two *terms* (as usual). She needed one that captures how experience itself can be rendered transcendental. She calls this "relation." Relation, for her, is the impetus of all productivity of social life. That impetus owes to all relations' genderedness. It is because gender is the principle of disguise, or "encompassment," thanks to which the empirical reality is never what it seems as it is given in its *terms*. Her "gender" *is* "relation," as we now understand the latter as a "twoness" that renders experience transcendental – via self-dividing every *term* given to experience into two. The art of the productive act is about making this never self-apparent, gendered reality episodically legible.

Strathern's relation, then, is neither an object of empirical observation nor countable. Hence the epigraph – "one is too few but two are too many." *The Gender of the Gift*, she says, is written against the trend wherein "anthropologists . . . have been encouraged to think [that] the alternative to one is many" (GG 52). The well-known bias for the individual person is only a subset of this partiality towards multiplicity that is but a unity of discrete parts (which are themselves exterior to each other). Strathern's relation, however, as an alternative to the discrete, does not follow the intuitive course of a solution: replacing the bias for the discrete is not performativity, read as processual construction in time. Rather, relation, for her, precedes the conceptual alternative givenness/construction, thing/performance, or nature/culture. That is why relation in Strathern's work is primarily about what she calls objects – or self-dividing objects to be more precise. Relations are aimed at objects and result in objects. We may call her objects multiplicity in its full transcendental or "intensive" sense.[17] Persons are one kind of such objects. As such, Strathern's "gender" is not reducible to beliefs or worldviews – qua conceptual counterpoints to the biologically inflected notion of "sex."[18] It is closer to a kinesthetic or "cinematographic" notion[19]: literally, nothing happens without it. She opens her account of gender as follows.

Across Melanesia, a person (human or non-human) is deemed constitutively dividual, only situationally individual.[20] Both states are different configurations of the universal condition of androgyny. The androgyny of a person manifests itself in two ways. First, when a person is viewed as part of a gender-unspecific collectivity – the "one-is-many mode" – each person is seen to encompass *within* themselves an androgynous, composite identity. This state is called "same-sex." The same-sex figure still has parts or appendages "belonging" to the opposite sex (GG 122). Second, when a person is gendered – the "dual mode" – the androgynous constituents within are thought to have been *externalized*, in order to privilege one of the halves – male or female. This state is called "cross-sex." A person is cross-sex when s/he is posited against another person with an already-externalized gender determination. That is, an externalized "male" or "female" person triggers the opposite in another. Whether a person is the cause or the effect of a trigger is always an ambiguous affair – due to, as we shall see, the "recursive" causality. In Strathern's succinct formulation, we get a case of cross-sex when "*what was 'half' a person becomes 'one' of a pair.*" In all, then, there are "*two forms of plurality (the composite and the dual), so there are two forms of androgyne*" (GG 15; emphasis added).[21]

There are no two selfsame sexed *terms* here. Twoness is rather in the two different ways the would-be sexual subject manifests the indeterminate, transcendental substance here called androgyny. Androgyny can be either external or internal in relation to the subject. When external, androgyny "contains" the subject as one of its constituents, either feminine or masculine. When internal, androgyny is "being contained" within the subject, making the subject's sex undecidable (between the feminine and the masculine). What is crucial is that the "protagonist" of this story is neither the feminine nor the masculine. It is a sexually indeterminate (some-)"thing" called androgyny. Thus, it is incorrect to say that androgyny here is made of both sexes – as the proverbial usage has it. We cannot say what androgyny *is* – because the term only positivizes relationality as such. We can say only that androgyny elicits sexual states in the subject depending on whether the former is external or internal. One might say that "androgyny" here designates multiplicity that is both "too small" and "too big" vis-à-vis the unity of a person.[22] A sexed or gendered body, then, must be something that is always "more" or "less" than as it is given to experience.

In Melanesia, any productive relation is thought to result in an object, be it a child or yams growing in the garden. Objects are productive relations successfully made visible. This covers everything "from the manner in which crops are regarded as growing in the soil to a dichotomy between political and domestic domains" (GG 14). Successes vary and are subject to aesthetic judgment, because objectification is not automatic instantiation of a known state (GG 174). Nor is it a natural extension of a person's endowment. It is a theater of improvisation that must negotiate with an inherent ambiguity and duplicity of an object. The idiom of that duplicity in object is gender. The negotiation between the two gendered moments – androgyny-within and androgyny-without – makes relationality an art of "secrecy and exposure" (GG 190).

The fact that two axes of a variable are simultaneously at work in a gendered object – i.e., male/female and container/contained – complicates the causal relation in objectification. A relational transformation – say, a girl growing into a woman – is not a contingent achievement or construction in processual chronological time. We tend to think so because we privilege the human actor, whose action is thought to demarcate before/after. The Melanesian emphasis is on the object – with multiple axes of variables at its disposal to effect a transformation. Its workings are counterintuitive. "Objects are apprehended as both cause and effect of the relations" (GG 221). The outcome is already in the object. A girl who is going to grow into a woman already has that potential within her. It is as though, just like androgyny, the "cause" is both internal and external to her, simultaneously working within her *and* awaiting her as a would-be effect. "The girl about to be impregnated already has a child within her" (GG 220).[23] She "recollects forward," as if she has "attained the point immediately preceding [the mother and child's] respective differentiation."[24] Strathern calls this aspect the "recursiveness of the anticipated outcome" (GG 221).[25] The recursive causality owes to object-centeredness, objects which are gendered. Gendered, they are internally divided by androgyny. All "particles" are already immanently at work – "given" like "nature" – for an "anticipated outcome," so to speak.[26] A contingent manifestation is merely a "rearranging" via internalization or externalization. Androgyny here is a multiplicity that destabilizes the container/contained distinction. Productivity measured by these two axes results in an interesting image. In place of processual performative achievement in time, it resembles an inside/outside *overturning*. The non-processual temporality is such that one is never a positively gendered being in the present but is always "too early" or "too late" for it. In this sense, gender not only disguises and renders the reality more than meets the eye. A Freudian drive, it "moves" everything.

repeating the impossible

Let us consider an example. Traditionally in anthropology much attention is given to male initiation ceremonies as a decisive moment for instituting the male difference. In Melanesia, Strathern notes, such is hardly the case. The Gimi initiation ceremony, for example, does not "construct" a male child into a man. The child does not get endowed with some gender-specific attributes. Rather, "he learns that what he already has (his genitals) are signs of encounters with women that already have taken place" (GG 112). Implications are far-reaching. That paradigmatic male organ only "commemorates" the fact of men's "theft" of power from women from time immemorial (GG 112).[27] The penis is commonly associated with the womb across Melanesia (GG 112, 260). It gets the central attention in male initiation ceremonies not because it is always already paradigmatically male. It is rather because of its ambiguity, self-division, and bisexuality – that is, because of its objectal impetus towards metamorphosis.

A ritual with an object (i.e., the penis) as its central agent, the Gimi initiation does not celebrate the origin of manhood, or, more precisely, the human agent's power to originate. Rather, "commemoration" replaces origin. A commemoration is a (mere) *repetition* of what always-already happened. What happened, furthermore, is not an earlier origination, but "theft." The penis here is "given" as an "ontological debt" – and not constructed. Yet neither filiation nor identity reckons its emergence. If anything, what is "commemorated" is an "interference" of filiation[28] – a theft. The fact of theft announces a "closure"[29] of anticipated continuation *pace* "still one more" (of the rule).[30] If we still call it a filiation, it is a mythical kind that happened in a deterritorialized time of "antimemory."[31] Here, "anything one has acquired . . . appears to have its origin elsewhere" (GG 221). "Transcendental memory," writes Deleuze, "grasps that which from the outset can only be recalled, even the first time." What is thus recalled, he continues, is "not a contingent past, but the being of the past as such," an "*a priori* past."[32]

We may call the male organ in the initiation ceremony a virtual object, after Deleuze. "[Virtual objects] are planted in [the world] like trees from another world, like Gogol's nose or Deucalion's stone." That is why the phallus, he writes, is "impossible," for it is "found only as lost." "The symbol," he goes on, "is the always-displaced fragment, standing for a past which was never present."[33]

Strathern has concentrated on objects in order to thematize something that is both "given" and "constructed" – "nature" and "culture." Something that is given yet requires a construction again can only be one thing: a lost object. It is an object whose "enjoyment" in the present is constitutively barred, made impossible. The initiation ceremony is, then, a technique of "retrospectively" constituting an "object of satisfaction" in the Freudian sense – precisely via making it constitutively lost (see Copjec, present issue). When thus positing a thing that is both "given" and "constructed," Strathern is not seeking some kind of theoretical compromise, a conceptual "middle ground." Rather, she is trying to translate for us the miraculous power in the relationally objectified thing (invoked by the "act" of a ceremony) as the Melanesians see this. Miraculous, once again, in Georges Bataille's full sense: "impossible and yet there it is."[34] In that sense, I submit that Strathern's objects are both impossible (*pace* Lacan) and virtual (*pace* Deleuze). They inscribe the miraculous feat of affirming the impossible as such.

In that "impossible" and deterritorialized dimension of "equivocation" (as Viveiros de Castro writes) where "the rule does not allow to traverse" (as Badiou writes),[35] what moves is not assigned difference mediated by the rules of convention but difference by repetition alone. Repetition, in turn, is realized in objects alone. An object is such an agent because it is gendered. Gendered, it is internally divided (by the androgynous substance), primed for inside/outside overturning that traverses across the "impossible." Viveiros de Castro's multinatural "body" is precisely such an object.

multinaturalism's body

It is clear, then, that for Strathern as well as for Viveiros de Castro the body is not the *factum brutum* of inanimate substance that subtends uninaturalism, and the omni-comparativism it in turn supports. When Viveiros de Castro speaks of the difference in "perspectives" that is "located in the bodily difference between species," that "body" is *not* an axis of a species-determined relativistic empirical world. It does not demarcate a unitary world of species-specific phenomenology that is but one contingent and particular instant within the universal totality of species variety. Such an Aristotelian view would perfectly agree with the multiculturalist aesthetic. Both conceive the body as a thing that is (not lost but) enjoyable *in the present*.

Multinaturalism's body cannot be more different. As a "vehicle" that traverses the impossible, that body strictly belongs to the aforementioned a priori temporality of repetition. As such, it is not a thing enjoyable in the present. The embodied species relation between two different species, for instance, is not a dyadic opposition that is mediated by a "third" substance of a universal compatibility we call nature.

Put this way, the dimension of impossibility that multinaturalism's body inscribes is identical to Lacan's impossibility apropos sexual relation. For Lacan's problematic is also that of doing away with a tacitly implied "third," which mediates sexual difference to make it a dyadic opposition-in-compatibility. His question, then, is: how does one conceptualize sexual difference from the immanently situated "perspective" (in Viveiros de Castro's sense) of either of the two, without posting a neutral third vantage point? Lacan's answer is that, immanently, sexual relation is neither negative nor positive, neither absent nor present. We simply cannot determine it. All we can do is to *positivize* indeterminacy itself. How? We declare that sexual relation is that peculiar kind of relationality that is made possible by the "experience" of impossibility as such. Sexual relation is a paradoxical kind of relatedness that is possible only in the form of impossibility. This statement merely affirms that "some" thing exists – that it is not nothing – but as to the quality or content of that thing we cannot determine (for immanently situated within it). That is why Lacan's proposition "there is no sexual relationship" should not stand by itself; it needs to be supplemented by "there is (some) One."[36] That "(some) One" affirms impossibility as such. That (some) One is a constitutively lost object of enjoyment. And just as in Strathern's androgyny, sex designates two ways to relate to this (some) One. We might call the latter singularity with two "uneven sides."[37]

In place of Lacan's (some) One, Viveiros de Castro posits what he calls the body. And where there is body, there is "perspective." What does that mean? Surely, we do not mean that a perspective is confined within the *factum brutum* of a species-specific body. As such, perspective is not something that comes "between" my body and that of the other species. Perspective rather attests to my inability to coincide with myself because there is the body (whether mine or the other's). As in Strathern, the body is the protagonist here: the body, like Lacan's (some) One or Strathern's androgyny, is that in relation to which I acquire a perspective. But let me pause here. In order to further align Strathern and Viveiros de Castro with Lacan we need to spell out the idea of limit implicit in Lacan, which is Kantian.

According to Deleuze, with Kant the idea of limit has changed. Before Kant, limit was only external, a fixed point that can be transgressed.[38] The Kantian limit is both internal and external – both in me and confronts me from outside of me. It is because the notion of limit also began to include the "passage to limit." The limit for Kant is "no longer the operation which limits something, it is on the contrary the term towards which something tends." The idea of limit now encompasses both the "tendency and that towards which it tends." In other words, it is both a tendency "in me" and that which makes me tend (from outside of me). Moreover, limit is now both the "passage" towards an Other and also that Other. In relation to such a limit, then, the Kantian subject is divided in between two different temporal points. The subject is, so to speak, a "cinematographic" effect in between these two "moments."[39]

The body that induces perspective is such a limit that is both external and internal, both to which one tends and the "act" of tending itself. By "act" I mean a self-referential fissuring into two of myself both as the agent of that "tending" and the patient who is being so tended. What Kant has captured is a slight temporal gap in every self-referential fissuring-into-two vis-à-vis limit – I am simultaneously "here," tending towards the limit, and also "there" with the limit. This, I think, is the reason why the body for Viveiros de Castro needs to be supplemented with "perspective." The presence of perspective signals the body's temporal displacement from itself. It is thanks to this gap that different species "meet" at the limit (read, the body) – "pronominally," as though referring to oneself as other.[40] As such, the multinaturalist perspectivism is not at all a bio-relativistic skepticism.

Once we factor in what we are calling "act" here, it makes perfect sense that Amazonians view inter-species relation as "alliance." Non-humans are affines to humans, for example. Such is not a metaphor, Viveiros de Castro emphasizes.[41] Just as in human affinal relations, inter-species relation is forged "counter-naturally" by alliance alone. Forged counter-naturally because an alliance is never prescribed by descent.[42] There is no "third" qua kinship, culture, or language that mediates two given allying groups. Nor is alliance merely a performative feat of intersubjective negotiation that "invents" relations – as has been observed repeatedly.[43]

Viveiros de Castro illustrates the inter-species alliance by way of Deleuze and Guattari's famed example

> of the wasp and the orchid, an assemblage "from which no wasp-orchid can ever descend" – and without which, one might add, no

> wasp and no orchid as we know them could ever leave descendants. The wasp-orchid assemblage is the origin myth of both the wasp and the orchid tribes: their shared deterritorialization, their common antimemory.[44]

This alliance assumes neither anterior filiation nor posterior descendants. There is no given continuous symmetry between "before" and "after" that an alliance passively receives. Rather, alliance is an act that installs a caesura in time; it installs its own "before" and "after." It is not merely a creative construction in processual time; not only the future but the past as well changes by an alliance. A "counter-invention,"[45] it is an act of constructing something as given always-already. A "discourse of the Given," it "sets the terms of the ontological debt."[46] An alliance "recollects forward" recursively to make something that lacked presence come into view. And as has been mentioned per recursivity, something that is putatively given yet requiring construction again cannot but be a lost object. That is why "myth" is invoked in the quote. "Antimemory," mythical time is deterritorialized, and is outside the chronological time of lived or livable presents.[47] The wasp and the orchid do not encounter each other in the present; they only meet at the limit.[48]

We now know that when we speak of "meeting at the limit" we have to posit (some) One – that is reducible neither to the wasp nor to the orchid, so to speak. The impossibility of relationship (in the mode of the subject–object encounter) must be affirmed in the presence of (some) One. It is such a positivized non-relationality – what Deleuze calls "disjunctive synthesis" – that Strathern calls androgyny. It is necessarily a "post-social" concept.

A striking resemblance to Lacan's theory of sexuality cannot be missed. Strathern expropriates twoness from predicating man–woman union; Lacan announces sexual relation qua harmonious union of two as impossible. Strathern does it by supplementing man–woman union with androgyny; Lacan supplements the thesis of the impossibility of sexual rapport with that of "there is (some) One."[49] The Lacanian technical term, "supernumerary One," is fitting. For Strathern's androgyny is neither subject to enumeration nor experience – it is always undecidable as to whether a given person contains androgyny or is being contained by it as one of its constituents.

the miracle of the socius, according to anthropology

Assigning a positive role, if not positivity, to supernumerary "post-social" entities has been anthropology's impulse, if not coherently realized practice. It was due to the prevalence of "dividual" or "fractal" personhood in non-Western societies, as Strathern has shown us. Claude Lévi-Strauss was arguably the most influential in furthering this impulse. Strathern is one of the most abiding heirs to this facet of Lévi-Strauss. "Facet" because Lévi-Strauss's work tends to be divided: on the one hand, there is a radical destabilizing of the human social; on the other, there is his structuralist answer that reinstates all the exchangeist and functionalist assumptions of the human social – whereby relation is treated as "molar logical objects, apprehended essentially in extension, as oppositions, contradiction and mediations."[50] *The Elementary Structures of Kinship* is a typical example of that tendency. The work begins with a radical reorienting of the central anthropological debate of the time: where does nature stop and culture begin? Lévi-Strauss turns to the phenomenon of incest prohibition to address this question. It is because incest prohibition occurs universally in every society, and yet the "content" varies from society to society. In incest prohibition he has found the "fold" between nature and culture.[51]

Lévi-Strauss was well aware that the presupposed either/or dyadism was the obstacle. Thus, approaching this division immanently, he conceptualized the boundary as a limit – in the precise sense we have established above. For Lévi-Strauss, however much we take apart the prohibition, we cannot discern the point of encounter between "nature" and "culture." It is because the prohibition's *raison d'être* is the "working out" of this very problem. This *act* or "passage" (towards the limit) of prohibition is what makes us cultural beings – not because humans, unlike non-human animals, possess the universal aversion towards so-called incest. Rather, "culture" *is* this self-fissuring act, "tending" towards the fold of a "double-headed thing with two unequal or uneven 'halves.'"[52] Hence this enigmatic expression: "the incest prohibition is at once on the threshold of culture, in culture, and in one sense . . . culture itself."[53] To paraphrase Lévi-Strauss, we are "human" because we have the incest prohibition.[54]

This breakthrough in rethinking dyadism by Lévi-Strauss is given an explicit articulation in *Anti-Oedipus* as a case of a *limit* concept – although the credit is given mostly to Marcel Griaule. Incest prohibition, as Deleuze and Guattari explain, is *not* a response to an actually occurring behavior by that name.[55] Prohibition does not come "after" what was once possible, and then prohibitively causes its impossibility: there is no "ready-made" object of incest prior to prohibition. A Kantian limit, incest prohibition punctuates a caesura in time, which itself distributes "before" and "after." This means that the act of prohibition makes its own referent. It does so not in the contingent time of "construction," of course, but "recursively," as though "recollecting forward." This implies that even as the prohibition makes its own referent, it simultaneously "disfigures" it.[56] For the caesura of time can be signaled only if incest is "constructed" precisely as a category that had never existed as such.[57] Incest prohibition for Deleuze and Guattari is a limit concept in this strange sense: it *prohibits* what is simultaneously avowed as always-already impossible. But how exactly is it done?

The definition of incest is such that for an empirical incest to be realized, the person (say, the biological sister) and kinship name (the kinship name "sister") must coincide in the sexual partner. The prohibition recursively "pre-empts" (or "recollects forward") this coincidence via "dishonoring the guilty.[58] "Dishonoring" involves a re-constituting of the guilty as one altogether and always-already outside the society of the desired/desiring – not just as an ex-insider who "got caught."[59] That is to say that incest prohibition's *example* is made not from "inside" but every time from "outside," only as an *exception*.[60] This is to repeat that per the act of disfigurement there is no neutral perspective from where the "before" and "after" states of the prohibited thing can be evaluated. The prohibition itself is that "differentiator."[61] It introduces a "breach" in causality.[62] Incest prohibition, then, is none other than a positivization of impossibility as such. It is a technique of recursively constituting the "lost object" of the erogenic category that is incest.

Now, it doesn't seem like a stretch to argue that incest prohibition works just like Melanesian androgyny. Neither "cultural" prescription nor "natural" substance, androgyny is an enigmatic "object" that is supplementarily posited to suspend the numerical determination of a sexed person at a given moment. So kept alive productively are the problems "two-in-one or one-in-two?," "containing or being-contained?" – whose effected sentiment will read, "one is too few but two are too many" (see epigraph). For Strathern, it is this undecidability that invites productive performative "movement" as individuals strive to will. But "will" as one may, it is, if I may repeat, only to confirm that "all the particles are already there."[63] It is, once again, a kind of "construction" which only affirms the "givenness" of things; an exception, shall we say, that announces itself only as an example. Temporality here is both "yet-be-be-completed" and "already-completed." Strathern calls such a point of a "fold" between exception and example *act*.[64]

impossible and prohibited

Now, with the aid of the Melanesian/Amazonian theses, I will try to illuminate Lacan's opaque "formulas of sexuation." As the subtitle of his Seminar XX reveals, Lacan's theory of sexual difference concerns knowledge. As such, Joan Copjec's groundbreaking aligning of Lacan's said formulas with Kant's antinomies is thoroughly justified. I will follow Copjec's lead here while inflecting her thesis with the foregoing discussion, especially through Viveiros de Castro's idea of "perspective."

Perspective, as we have established, becomes relevant because of the presence of limit, the Kantian limit to be precise. Kant's idea of limit introduces a certain temporal gap as the subject comes to occupy two different temporal "moments" at once. As was shown above, perspective is operationalized because of that gap. It seems to me that the idea of perspective plays the same role as that of knowledge in Lacan's formulas of sexuation. Knowledge is an issue in the latter because Lacan's formulas are not directly about how femininity and masculinity are in themselves; it is neither about what women and men are from a "third" vantage point; nor is it about how men and women capture the world or each other differently.[65] Both said views remain within the either/or logic, as though women and men are positive beings describable either "subjectively" or "objectively." Once the self-fissuring apropos knowledge or perspective is factored in, either/or is no more. Lacan's formulas are, then, about how we relate to the limit posed by the problematic of the other sex. There are two ways in which the subject's knowledge relates to the limit – one feminine, the other masculine. Insofar as the feminine and the masculine are *not* positivities, we do not have two different limits. Instead, the limit that causes knowledge's "failure" is singular. More precisely, it is a supernumerary (some) One – with two "uneven sides," so to speak.[66]

Seen this way – that is, through Viveiros de Castro's "perspective," and then through Lacan's formulas – Kant's antinomies can be also seen as strictly about limit that is singular. Again, I am thinking about Deleuze's articulation of Kantian limit. Deleuze broaches this issue (in the lectures on Kant) when he brings up Kant's idea of time. With Kant, he says, time "no longer limits the world"; "it will traverse it." Traverse, because time "drags (us) along."[67] It is as though time, traversing through me, never allowing me to be present to myself, does not just leave me behind, as it were. It also compels me forward as though along with it. That is why, as we have examined above, limit for Kant is also a "passage" to the limit. It is as if time, never allowing me to be present at myself, renders limit constitutively *impossible* (to reach). Yet time, dragging me along, does not let me stop pursuing "it." This is to say that Kantian time renders limit a lost object (of enjoyment). It is not only *impossible* but also *prohibited*. Prohibited because limit (as something external to me) not only forever escapes from me but I *make myself* fail to reach it, so to speak, precisely on account of ceaselessly pursuing it in spite of myself.

Allow me to repeat that this limit is both "internal" and "external." The impossible and the prohibited are not "in" the subject as anthropocentrically conceived. Put in a Deleuzian fashion, by subject we mean an effect forged in relation to limit qua singularity, whereby the "series" called impossibility and the "series" called prohibition meet.[68] The former, for Deleuze, is in the order of progression or "step by step"; while the latter is accessed "all at once."[69] We called the same yet-to-be-completed and already-completed. Note that "all at once" for Deleuze is also a kind of series. For singularity to appear, for him, *you need both series together*. For singularity is none other than the point where the two series "meet," without encountering. From the "perspective" of the subject who relates to the singularity, then, the two series – "step by step" and "all at once" – never appear independent of each other.[70] I think this is crucial in understanding Kant's antinomies.

I believe that it is from this angle – i.e., that you need both series together – that we should appreciate the Kantian idea of time. I mean that time for Kant is no more in the order of "extensive magnitude." Kantian time is not a "succession of parts," or "parts that are external to each other."[71] If so, the "infinite regress" pertaining to the mathematical antinomy (cum the feminine sex) should be understood in this non-extensive sense. That is to say that the mathematical antinomy's infinite regress is *not* a homogeneous series of "diachrony."[72] It is always-already inflected, as it were, through the "dynamical series" qua "all at once." Rather than the "still one more"[73] series of a simple addition, it is closer to infinite division or "infinitely doubled difference"[74] – which is simultaneously self-division insofar as limit is both internal and external. I wonder whether Lacan's feminine "not-all" is this kind of "impossibility." What I am contrasting myself to here is the interpretation of "not-all" as the "process" of a contingent (and open) series of encounters with irreducible particulars.[75]

The dynamical side can be similarly amended. It is not directly about "structure" – as it is often contrasted to the mathematical antinomy's "series." I am thinking of the interpretation where the so-called mathematical "series" is contrasted to the "dynamical" "structural" problematic of the undecidability between inclusion and exclusion. Rather – if we persist in the spirit of taking both aforementioned series into account at once – the "prohibitive," dynamical self-referencing is always-already inflected by the "impossible" dimension of Kantian time. As Deleuze puts it, any idea of structure needs two "series."[76]

Then, if seen from the angle of Kantian limit (as suggested by Deleuze), can we not say that Kant's antinomies are about this fissuring, doubled-up failure of knowledge – whose names of "attributes" are impossibility and prohibition?[77] Can we not align mathematical antinomy with the "impossibility" (of reaching the limit), and the dynamical with "prohibition" (qua self-prohibition called castration)?[78] Then, we should say more precisely that the twoness of sex does not indicate (in Lacan's formulas) two separate ways in which knowledge fails – which seems to, once again, bring us too close to dyadism. Rather, we should say that twoness corresponds to the two "uneven sides" of knowledge's failure vis-à-vis the singular limit qua the supernumerary (some) One. Only this way can we unambiguously distance ourselves from the temptation of assigning Lacan's twoness to two autonomous orientations – one possessed by woman and the other by man. Only then can we say: all there is that there is that (some) One we are compelled to posit without being able to wrap our minds around it; that what we deem sexual difference is a folk explanation, a prosthetic response that compensates for this constitutive failure of knowledge.[79]

set theory

I believe that the above reading of antinomies better aligns Lacan's formulas with the Cantorian set theory on which he relies. For Georg Cantor's theory is that of the One, not of the two. The Melanesian thesis helps here. Likewise, Strathern's gender theory is about singular androgyny's two "attributes" – androgyny-without and androgyny-within. It is just as Lacan's formulas, *pace* antinomies, are about two ways we relate to the singular limit. It is worth repeating that the Melanesian thesis is *not* about some self-same exceptional One that manifests itself in two examples, cross-sex and same-sex. It is rather about two oscillating forms (i.e., androgyne-within and androgyne-without) which unsettle the very distinction between exception and example, or between containing and being-contained. An equivalent unifying theory in Lacan is set theory, which is also about the undecidability of inclusion/exclusion.[80]

Cantor's set theory is said to have conceptualized the "real infinity."[81] Previous notions were confined to "one-infinity" or the "unicity of infinity." One-infinity because totality was conceivable only on account of positing an exception to it, a transcendent exception that is hierarchically represented as beyond. It was a domesticated idea of infinity with a "center" that provided unicity. Our usual idea of "something that goes on and on" (that we may now call the extensive infinite) is of this variety.[82] Cantor's set theory achieved a de-domestication of sorts by abolishing the theretofore tacitly posited center. The rightful multiplicity is thereby conferred to infinity, replacing the secretly unified kind. He has done this by identifying an inherent paradox attendant to any reflection on totality, that is to say, the set of all sets.[83] The paradox is the following: is that set one of its members, or not? Insofar as this set claims to be "all," it surely must include itself; but if it does, it is not "all" but only one of its members. The paradox lies in the fact that such a set is both included and excluded from its own claimed totality – hence, paradoxically All and not-All, an exception to itself while simultaneously an example of itself.[84] In this sense, set theory for Lacan is an effective theory of One, of One that can never coincide with itself. For, again, the very gesture of claiming to be One makes it stop being One. The reason for this non-coincidence we can call *self-referentiality*.[85] However, per our discussion above, we need to be careful not to identify this singular point of self-referentiality as "structure," as opposed to "series." Only in so doing can we say that Cantor's theory accommodates both of Kant's antinomies.

As Roy Wagner puts it, "Cantor used the intuition of infinity *in reverse*."[86] That way, Cantor dealt with number "from the 'outside,' that is, in terms of its conceptual boundaries."[87] What is the significance? Two different "series" are simultaneously taken into account in this operation, resulting in addressing both of Kant's antinomies. The first is the potential infinite, that is, totality as progressively or regressively approached. Cantor supplements this with a second series: totality considered "all at once," as though it is "already here." It is this claim (of "already here") that invites the paradox of self-referentiality. The two series roughly correspond to Kant's mathematical and dynamic antinomies. In the former, infinity is "impossible," and in the latter, (self-referentially) "prohibited."

In this way, Cantor is said to have "ontologized" infinity, making it expressible by number.[88] Ontologized, infinity makes its appearance in Cantor's work as a "determinate object" (II 111). And this object is just like Strathern's object.[89] Gendered, as I write above, her object is the constructive impetus of the socius par excellence. But as I also note, the gendered construction only reveals that "all the particles are already there." That is to say that Strathern's object is both something *yet-to-be-completed* and *already-completed*. Her own term for this double temporality is *recursivity*. Strathern's recursivity captures none other than how an object is turned into a "lost object," and is thus magical.

infinity, object, gender

Let us recall that the issue of infinity was raised at the outset in order to offset multiculturalist *pluralism*. What we learn by putting Strathern and Cantor side-by-side is that infinity calls for the object. And that object, whether seen through Strathern or Cantor (made Kantian/Lacanian), is gendered. This course of reasoning, then, reveals the problem with pluralism: pluralism makes the "magic" of the object disappear, because of its ambition for a de-gendered view of things.

We come to this conclusion not because the theories of Lacan et al. are "applicable" to Melanesia or Amazonia. There must be another compelling reason why Melanesians, like good Lacanians, go against the intuitive accrediting of societal productivity to the union between the tangible woman and the tangible man. From where does this impulse of "transcendental empiricism" come? This reasoning comprises anthropology's foundational post-social insight: the "magic" of the socius has to be at once "given" yet "impossible," which can only be inscribed in the gendered object.

Strathern's most radical contribution lies in showing convincingly that without gender that anthropological foundational insight remains only theoretical; it would not know what it is to encounter the social as a miraculous object. *Without gender, there is no such a thing as anthropology*. Some twenty-five years after the publication of her seminal ethnography, that forceful feminist intervention now sadly needs to be rediscovered. It is largely lost in mainstream disciplinary anthropology. Our inability to read psychoanalysis and vitalism side-by-side, I will argue, is symptomatic of a related broader trend. Losing sight of the object and gender resulted in a kind of return to the multicultural-particularist omni-comparativism to which Strathern had reacted. This has only accelerated anthropology's growing irrelevance to theory making, which never volunteered cross-cultural comparativism in any case.

the "intimated" infinity

Let me comparatively discuss one of Strathern's contemporary ethnographies on the same region as an example of the de-gendering trend: the already-mentioned Jadran Mimica's *Intimations of Infinity: The Mythopoeia of the Iqwaye Counting System and Number*. De-gendered, I will show, infinity turns into plurality.

The Iqwaye count with fingers, toes, and the body. Their counting system, as Mimica emphasizes, cannot be separated from these "number-objects" made of the human body and its parts (II 13). It is so much so that Mimica can state: "numbers are human beings" (II 72); or, to be more precise, "being of number is the corporeal man" (II 104). Much is at stake in what he means by "object," which, in turn, determines what he means by "intimation." (Why is the Iqwaye infinity intimated? Why can't it be, say, a "theory" just as Cantor's is?) This word choice, as I will show, has implications for how we envision West/non-West comparativism.

An Iqwaye person begins to count with the thumb of either hand digit by digit (1–5), which is continued on the other hand (6–10), then to the toes (11–15 and 16–20 respectively per foot). In sum, one person stands for 20. For this reason, 20 provides the most significant sense of unity, especially when dealing with large numbers. Hence, 20 is represented, retrogressively, once again, by a digit. Thus, 20 digits can represent 20 men, which results in 400 ($20 \times 20 = 400$). That is to say that numbers larger than 20 proceed in the fashion of a series: 20, 20^2, 20^3, 20^4 and so on. Predictably, 400 is another number with a sense of unity, and is, again, retrogressively represented by one person, analogously to when 20 is represented by one digit (II 41).

Characteristically, then, the Iqwaye counting periodically retrogresses into a starting point, whether one digit or one person. In counting, one might say, "the whole is equivalent to its part" (II 52–53). Mimica calls this reigning figure "homology," and deems it the central point of his observation. A homologizing operation, then, counting here is a "transformation of multiplicity into singularity or of the many into oneness." A "totalization" is immediately also a "particularization" (II 56). Somewhat haphazardly, I described the same at the outset as a simultaneously "diverging" and "converging" counting system.

Given this, Mimica's query on the Iqwaye understanding of infinity seems to be a natural step: how would a people whose counting regularly returns to the starting point, so to speak, conceptualize something like infinity? To answer, Mimica turns to their myth. My first concern here, though, is with what he means by "intimation" and the notion of "name-object" that eventuates such recourse. The richest clue we find is the chapter in which he contrasts the Iqwaye system to the conventions of Western mathematics.

For Mimica, the Iqwaye idea of number is inseparable from the "life-world" or the "emotional meanings of the human placedness" that the counting practice enacts (II 97). This "affective quality" related to number, for him, makes the Iqwaye system irreducibly different from the formal-abstract number categories of Western mathematics. This is not just because the Iqwaye count with the human body and its parts. Rather, to count in the lived world of the Iqwaye is also to live directly their "cosmology." Reigning over both, whether it is everyday counting or the mythopoetic cosmology, is the part/whole homology. Just as the greatest possible number the Iqwaye count is one, the individual person and his/her body immediately and directly partakes of the entire cosmic *substance* (my word). "One single finger," he writes, "like one body are already symbolic equivalents of the entire cosmos." This homologically embodied expression is what Mimica means by "intimation." "One by itself is already the intimation of infinity," he writes (II 95). It is a word intended to impart a "pregnant sense of infinity" (II 100).

Not surprisingly, within the Western mathematical tradition, it is Georg Cantor who receives Mimica's privileged attention. To Mimica, Cantor was an exceptional case of someone who had inherited the theological notion of "actual or absolute infinite . . . [that which is] given all at once" (II 107). His interpretation of Cantor will reveal for us how Mimica's "object" differs from that of Strathern in instructive ways.

As Mimica summarizes, Cantor has shown us what happens to numbers when they go beyond finite numbers: it becomes impossible to determine the part/whole relation between one set of "transfinite" numbers and another set – say, the set of infinite natural numbers 0, 1, 2, 3, 4, . . . and another set of infinite even numbers, 0, 2, 4, 6, . . . Clearly, the latter is a "subset" of the former in that all elements of the latter are also elements of the former (II 108–09).

However, both sets being infinite, they must also be equal in size in another sense. Hence, the doubling nature of the part/whole relation: an infinite set's "regions" also partake of infinity, allowing us to speak of different degrees of infinity within infinity (II 108). In this fashion, Mimica concludes, Cantor has devised ways to treat infinity as a property of sets, fully "structured," paving the way for so-called "transfinite arithmetics" (II 111). The infinite now becomes "transformed into mathematical and hence precisely *determined objects*" (II 111; emphasis added).

In this "determinate objectness," the affinity between Cantor's theory and the Iqwaye system is unmistakable. Nonetheless, or precisely because of this affinity, Mimica's criticism of Cantor is unsparing. He laments that in Cantor, "infinity is no longer a pregnant notion" (II 111). All Cantor offered was an "abstract concreteness." A product of "logico-mathematical reasoning," all Cantor's theory does is render "otherwise nebulous infinite" a "rationally manageable construct" (II 112). For Mimica, Cantor's intervention is an undiluted continuation of the Western mathematical tradition born of metaphysics, but with a pretense to be otherwise. As such, it "subsists through refractions . . . of its own illusory self-image," which is the "the self-sufficiency and autogenesis of its concepts" (II 112). Being merely "axiomatic," it is a "knowledge enclosed within itself" (II 107). "For Cantor," he concludes revealingly, "mathematics was primarily an articulation of relations immanent *in the mind*" (II 113; emphasis added). Being of the "mind," nothing could be further from the "life-world"-centrality of the Iqwaye.

Mimica's Cantor is quite different from Cantor made Lacanian above. In the Lacanian assessment, the relevant conceptual oppositions in Cantor lie elsewhere than mind/life. If anything, for Lacanians, Cantor's importance lies precisely in showing an effective way out of the enduring mind/body dualism. The key, as we saw, was the undecidability between inclusion and exclusion, between a set and its subsets; how any claim to infinity must also appear as one of its members. Put plainly, that is how a "claim" materializes itself as one of the "claimed" things. We have here an immediate collapsing of infinity with finitude, exception with example, or, if you will, "mind" with "object." In Lacanese, in Cantor we have a "*coincidence of signification and object.*"[90]

What Strathern means by "object" can be accessed in this Lacanian way, not via Mimica's homology. For Strathern's object is "miraculous." Miraculous because the social "thing" is a self-authorizing, self-referencing mechanism which not only signifies the world but signifies itself precisely as one of the signified "objects" – or the "body" as Viveiros de Castro has it. "Body" because it constructs itself as given – as though always-already given from the "anti-memory" time of "theft." Such an "object," foundational to anthropology, is emphatically not conceptually opposed to the "mind."

recursivity vs. homology

The crux of the matter is that what we saw as undecidability qua self-referential act, Mimica sees as a structural part/whole homology. Homology, for him, is operationalized by the shared cosmic substance. Positively – not "impossibly" – does the cosmic substance warrant the connectivity of opposites, be it one digit with infinity or the everyday existence with cosmogony.

Recall again that an immanent temporality, not some positive substance (which may be encountered in the present), was at the center of Strathern's thinking. Recursivity was her term. In the afterword to Mimica's very book, Roy Wagner employs the same term. Wagner uses the term to describe Cantor's innovation and its possible affinity to the Iqwaye system. Recall Wagner's characterization: he speaks of a "reversal" in Cantor. For Wagner, Cantor treated infinity as though it is already given. Cantor does it without bothering to settle the question of infinity's "constructability" within the extensive temporality. With a "boundary" and the "outside," we saw Wagner comment, Cantor's infinity suddenly – "impossibly" – acts like an object.

Clearly, that is not the kind of object Mimica was expecting – the betrayal of whose expectation made him declare that Cantor only dealt with things of the "mind." We can reword the subtlety of the Cantorian object in an antinomic way. Cantor puts together two propositions. First, there is the pre-Cantorian proposition that infinity is "impossible." To that proposition of "skepticism," Cantor merely supplemented another proposition that affirms our position of immanence: you cannot utter the first proposition (without committing the paradox of claiming "all") – the proposition is "prohibited."[91] Cantor's object is none other than this affirmed or positivized unsayable. It is an "object" only as much as a fault-line would be one (Copjec, present issue). One conjectures that it is this positivization of tension itself that Mimica disapproves of in Cantor's work.[92]

Myths and cosmologies are Mimica's alternatives to Cantor. They are offered as a mediating resolution to the Cantorian tension just mentioned. It is also telling that it is only in his chapter on myth and cosmology that the issue of gender gets addressed. Mimica demonstrates a stringent accord between the Iqwaye cosmology, on the one hand, and its kinship/gender practice and counting system, on the other. All three, for him, demonstrate how the problematic of twoness resolves itself by a fusion into One. In the interest of space, I will only discuss cosmology.

"Omalyce," the male primordial being, is said to have created the world through self-division. From his self-castration – no less – came not only man/woman but sky/earth, before/after, and, of course, part/whole. The two of difference originated from Omalyce's one of self-sameness. Omalyce the male is said to be a man when seen from the front, and a woman from the back. This way, "the unity of sexual opposites is achieved in the male sex" (II 84–85). To be consistent, Omalyce's sons are direct physical extensions of Him, not a product of the union in difference. Men are self-birthing beings, in short (II 88).[93]

Hence, when Mimica speaks of "maleness as bisexuality" (II 89) he does not come close to Strathern's androgyny. Unlike the latter, his two-in-one cannot possibly (or "impossibly") be imminent to a spontaneous "overturn" into one-in-two. It is because his is a homology. Homology absents any awareness of the paradox of one – the paradox subsisting in any *claim* to one. This despite the fact that maleness in his exegesis is also doubled: it is both the "genus" (that unites sexual difference) *and* one of its own "species." (For maleness *is* bisexuality and also one of the constituents of the same bisexuality.) Even in self-division, Omalyce's immanent twoness has nothing to do with self-referentiality. "Act" is an entirely superfluous notion in Omalyce's self-realization. Neither is his "myth" in the realm of "antimemory," invoking relatedness that is constitutively "elsewhere" or "lost." Instead, his myth enjoys a (homological) substantial continuity with everyday practices (of kinship and counting).

In short, for him, gender difference's twoness has nothing to do with the self-dividing experience of a limit in an immanent temporality. Rather, gender's twoness, for him, is but two manifestations of the same transcendent substance, which can appear either as different or the same as itself.

"intimation" and its comparativism

Having said this, the reason for Mimica's interest in the Iqwaye notion of infinity suddenly seems elusive. Strathern's interest in the topic, on the other hand, seemed directly related to her interest in gender: gender's twoness qua self-referentiality points to "auto-affective" quality of the socius; it is how society makes itself magical. That is to say, put with Clifford Geertz, it is how society "interprets" itself.[94] In contrast, the Iqwaye infinity seems to lack such a general theoretical import, since, according to Mimica, it can only be "intimated" by the anthropologist.[95] Why would he be interested in the topic, then? He invalidates Cantor's thesis by pitting it against the Iqwaye, but he is silent about what we should think of the Iqwaye thesis itself. The "validity" of the Iqwaye thesis is politely left unsaid.

Allow me to conjecture: doesn't Mimica actually want to say that infinity as the Iqwaye conceive it does not, in fact, exist? Isn't he saying, as a good multiculturalist, that what the Iqwaye might think as infinity is actually plurality (of the infinitely progressing/diverging kind)? If I may indulge one more rhetorical question: isn't he also saying that Cantor came close to the Iqwaye because, put ironically, he only stayed in the realm of the "mind"?

Assumed here is another dichotomy: there are "life-worlds" (read non-West) and there are "theories" proper (read West). The anthropologist keeps vigilance over the either/or logic of connect/disconnect between the two: theories are deemed either "Euro-centered" and disconnected or (surprisingly) "applicable." The theoretical basis of omni-comparativism, that which negatively motivated Strathern's intervention, is complete here. For comparativism's productivity relies on the division of labor between life-worlds and theory-worlds. Needless to say, the life/theory division presupposes uninaturalism.

conclusion

Strathern would agree that uninaturalist assumptions have nurtured the enterprising career of the "anthropology of gender." And that was at the cost of anthropology which *is* gendered. She would also agree with Viveiros de Castro that a multinaturalist anthropology, which is fully gendered, cannot but be "post-social."[96] It is because the fact of sexual difference operationalizes "impossibility" in all things deemed paradigmatically social – e.g., "society," "culture," "nature," "language," "kinship," etc. When deemed gender neutral and actually One, these ones mediate, or contain the fissure of two. In contrast, affirming impossibility as such means that we only affirm (some) One – not the actual One. It means the affirming of "this-side that cannot be represented as such in the complex" – to use Deleuze and Guattari's descriptor for incest prohibition – because the whole "complex is an element derived from this this-side-of."[97] That is to say that, gendered, society does not exist; it only insists in ex-sistence.[98] What I am trying to suggest in conclusion is that Deleuze and Lacan are incontrovertibly joined at this limit-concept that is post-social. Whether you call it "envelope" or "compact," "orgiastic representation" or "erogenous zones," the two thinkers converge on the point that the limit cannot be represented.[99] It can only be objectified – with an object that inspires us to say, with Bataille, it is "impossible and yet there it is." As an object, the limit is beyond the either/or measure, beyond the propositional complementarity-in-negation. This includes a proposition such as "You Can't Have it Both Ways: Deleuze or Lacan."

Peter Hallward's pronouncement pairs up the two thinkers in a certain rapport. For his appraisal focuses on a common referent between the two; one might say a "third" referent, whose respective representations by the two thinkers evidence the alleged irreconcilability. This common referent is "the subject." Hallward writes: "Deleuze rejects the category of the subject for the same reason that Lacan embraced it."[100] One wonders when the reason for such a diagnosis as "for the same reason" comes to pass. How does one arrive at such a vantage point of parity – the point from where the "absolute union (or) absolute aggressiveness"[101] kind of judgment can be passed? What is the "subject" anyway that it brings to terms such an absolute opposition or exclusivism of the two?

For Hallward, the truth of Lacan's idea of the subject is "interpersonal relations." Lacan's reliance on speech is Hallward's evidence. For "speech is by definition a trans-subjective activity," that is, "relations with the other (human, dare one say)." For Hallward, Lacan's privileging of the password says it all. For a "password means nothing, other than the institution of a shared or socialized sphere of meaning itself" In short, the truth of the Lacanian subject is convention (my word), a "symbolic being-with."[102]

Hallward follows this description of Lacan with opposing qualities in Deleuze. In so doing Deleuze's own fighting words are mobilized against psychoanalysis: "when we bang against things"; "the immediate intensity of things"; "things themselves . . . (in their) 'natural' state"; "Reality"; "Immediate expression of being itself," etc.[103] This roster of fighting words gives us a glimpse of another either/or question lurking behind the inquisitional vehemence of Deleuze-or-Lacan: "nature" or "culture"?

The position from where one can ask that question, *excepted* from the exclusivism of two, we have been calling here uninaturalism. The all-importance given to the notion of the "subject" by Hallward, then, does not seem to be all that post-socially informed. To be immanent to that exclusivism of two, on the other hand, is to be an *exceptional example*. Such is the anthropologically inflected understanding of "limit" I have been arguing here.

One cannot but wonder what we stand to learn from the difference named "Deleuze"-vs.-"Lacan." What is in the name of an author? Hallward seems to articulate this difference convinced that a textual or thought's coherence-in-time can be summoned under the biographical name of an author without excess. By excess, I mean debt, loss, theft, gift, and so on – what Strathern summarily calls "relation"; I mean the "recursive" hold that the debts or gifts of Lévi-Strauss, Amazonians, or Melanesians (or psychoanalysis, for that matter) had on Deleuzianism and Lacanianism. What do we stand to learn from an articulation of the Deleuze/Lacan difference that relies heavily on Deleuze's own copious complaints against Lacan? Does Hallward not write, therefore, from within Deleuze's own self-referential rhetorical space – which is by default denied to Deleuze's predecessor, Lacan?[104]

The point is not about fairness or balance. It is about how we understand the historicity of an idea or text; how we understand the decisionist act of theorizing that is not reducible to construction or invention by the heroic author. In fact, Deleuze himself commented extensively on how Hallward's kind of "a-relational" approach to historicity operates – its rhetorical structure, shall we say. Relevant here is Deleuze's critique of how our memory tends to privilege the "active" differentiation or synthesis of time, at the expense of "passive" repetition – of the debt, shall we say. This when in fact the latter "grounds" the former.[105] The active synthesis is self-referentially doubled. For as we represent the past, we also silently represent the present present in that representation. Deleuze writes: "it is of the essence of representation not only to represent something but to represent its own representativity."[106] The outcome of this doubling, if I may skip to Deleuze's conclusion, is our tendency to conceive of time as made of a succession of presents, all lived and livable – objectively, neutrally, and extensively, so to speak.[107] It is because, if I may paraphrase, we misrecognize the "extra dimension"[108] of our own self-referential doubling as the "extra-" surfeit of reality, i.e., the "out there" quality of the *milieu*.[109] The milieu gives us the illusion that "there must be another time *in* which (this very operation of synthesis) can occur."[110] Perhaps, uni-nature is such a milieu, "in" which we are inclined to find ourselves, objectively and gender neutrally, without debt.

If that is the temporal "rhetorical structure" of representationalism generally, we can only imagine it becoming more entrenched in an autobiography. Yes, that includes Deleuze's own autobiographing against Lacan et al., which passes up the "passive synthesis" that grounds relationality. The polemical autobiographer forgets even the most fundamental of the structuralist semiological wisdom: the author tends to present him/herself as an inaugural witness to the virgin commencement of "mapping" between signifier and the signified, as if not given over a priori to the "excess of signifiers," whose signifieds one cannot "map" but must assume.[111] If that is not the rhetoric of uninaturalism, I don't know what is. On the other hand, what academic makes a clean getaway from the trappings of representationalism anyway?

notes

For my teachers of anthropology in writing, Joan Copjec and Marilyn Strathern. I am much indebted to my tireless co-editor Arun Saldanha. The opening quote is from Marilyn Strathern, *Partial Connections*, updated ed. (Walnut Creek, CA: Altamira, 2005) 36.

1 Jacques Lacan, *Book XX. Encore, 1972–1973: On Feminine Sexuality: The Limits of Love and Knowledge*, trans. Bruce Fink (New York: Norton, 1997).

2 See Rane Willerslev, *Soul Hunters: Hunting, Animism, and Personhood among the Siberian Yuqaghirs* (Berkeley and Los Angeles: U of California P, 2007).

3 Eduardo Viveiros de Castro, "Perspectival Anthropology and the Method of Controlled Equivocation," *Tipití: Journal of the Society for the*

Anthropology of Lowland South America 2.1 (2004) 6, available <http://digitalcommons.trinity.edu/tipiti/vol2/iss1/1/>. Subsequent page numbers are given in the text.

4 Judith Butler, *Gender Trouble: Feminism and the Subversion of Identity* (New York: Routledge, 1990).

5 John L. Austin, *How to Do Things with Words* (Cambridge, MA: Harvard UP, 1975).

6 Peter Hallward, "You Can't Have it Both Ways: Deleuze or Lacan" in *Deleuze and Psychoanalysis: Philosophical Essays on Deleuze's Debate with Psychoanalysis*, ed. Leen de Bolle (Leuven: Leuven UP, 2010).

7 Gilles Deleuze, *The Logic of Sense*, trans. Charles Stivale (New York: Columbia UP, 1990) 257–58.

8 Jadran Mimica, *Intimations of Infinity: The Mythopoeia of the Iqwaye Counting System and Number* (London: Berg, 1988). Henceforth cited as "II," with page numbers given in the text.

9 Kiarina Kordela, *$urplus: Spinoza, Lacan* (Albany: State U of New York P, 2007).

10 Georges Bataille, *The Accursed Share vols. II and III: The History of Eroticism and Sovereignty*, trans. Robert Hurley (New York: Zone, 1991) 201.

11 See Claude Lévi-Strauss, *The Elementary Structures of Kinship* (Boston: Beacon, 1969) chapters 1 and 2.

12 Marilyn Strathern, *The Gender of the Gift: Problems with Women and Problems with Society in Melanesia* (Berkeley: U of California P, 1988). Henceforth cited as "GG," with page numbers given in the text. Let me hasten to add that I am no expert on Melanesia. Although I am an anthropologist by profession my qualification for the matter at hand is only that of an interested amateur, or dilettante, if you will. I would like to borrow this occasion to thank my Melanesianist colleague David Lipset, who has endured my credulity with the patience of a true expert. Needless to say, the resulting shortcomings are all mine.

13 Strathern, *Gender of the Gift* 379. Both Strathern's and Mimica's primary research sites, Sambia and Iqwaye, belong to the same language group, Anga, which is highly suggestive of intermarriage. David Lipset, pers. comm.

14 Roy Wagner, *An Anthropology of the Subject: Holographic Worldview in New Guinea and its Meaning and Significance for the World of Anthropology* (Berkeley: U of California P, 2001) 227.

15 Idem, "Afterword: Some Comments on the Iqwaye Mathematic" in II.

16 Gilles Deleuze, *Difference and Repetition*, trans. Paul Patton (New York: Columbia UP, 1994).

17 Alain Badiou, *Being and Event*, trans. Oliver Feltham (London: Continuum, 2005); Deleuze, *Difference and Repetition*.

18 For this reason, I will use the words "sex" (or sexual) and "gender" interchangeably.

19 Gilles Deleuze, *Cinema I: The Movement-Image*, trans. Hugh Tomlinson and Robert Galeta (Minneapolis: U of Minnesota P, 1986).

20 In the interest of the flow of the argument, I address variations within Papua New Guinea only when very necessary.

21 Ibid.; emphasis added.

22 Gilles Deleuze, *Proust and Signs*, trans. R. Howard (London: Athlone, 2000) 80. Such "identity of extremes" is Deleuze's definition of infinity; see *Difference and Repetition* 42.

23 Ibid. 220. Growth generally "has a retrospective character to it. It is in anticipation of the separation of grown thing from the grower that the thing so grows, for it is only known to have done so after the event" (280). Growth belongs to only one kind of object-relation called "unmediated." The other one is "mediated," which includes gift exchange. However, at the root of Strathern's feminist innovation is the elimination of this long-held distinction between growth (read female and convention-driven) and exchange (read male and inventive). This distinction disappears when a person is also a kind of object, when growth is also viewed as an inventive objectification that exceeds convention.

24 Eduardo Viveiros de Castro, "Intensive Filiation and Demonic Alliance" in *Deleuzian Intersections: Science, Technology, Anthropology* (New York: Berghahn, 2010).

25 Ibid. 221.

26 By "particle" I have in mind Gilles Deleuze and Félix Guattari, *A Thousand Plateaus: Capitalism and Schizophrenia*, trans. Brian Massumi (Minneapolis: U of Minnesota P, 1987): e.g., "Sexuality proceeds by way of the becoming-woman of man and the

becoming-animal of the human: an emission of particles" (307). This is different from the atomic view of gender propagated in Gilles Deleuze and Félix Guattari, *Anti-Oedipus: Capitalism and Schizophrenia*, trans. Robert Hurley, Mark Seem, and Helen R. Lane (New York: Penguin, 2009) 69–70.

27 Ibid.; see also Donald Tuzin, *The Cassowary's Revenge: The Life and Death of Masculinity in a Papua New Guinea Society* (Chicago: U of Chicago P, 1997).

28 Cf. Deleuze, *Difference and Repetition* 73.

29 Ibid.

30 Badiou, *Being and Event* 147. I also have in mind here Badiou's passage on the limit ordinal. The limit ordinal is Other, in that "it can never be the still-one-more which succeeds an other." "Not belonging to itself," he continues, "it ex-sists from the sequence whose limit it is" (150).

31 Deleuze and Guattari, *Anti-Oedipus*.

32 Deleuze, *Difference and Repetition* 140, 81.

33 Ibid., 101, 103.

34 Bataille, *Accursed Share* 201.

35 Badiou, *Being and Event* 147.

36 Joan Copjec, "*Gai Savoir Sera*: The Science of Love and the Insolence of Chance" in *Alain Badiou: Philosophy and its Conditions* (Albany: State U of New York P, 2005) 123.

37 Deleuze, *Logic of Sense* 50.

38 Gilles Deleuze, unpublished lecture on Kant, 14 Mar. 1978, available <http://www.webdeleuze.com/php/texte.php?cle=66&groupe=Kant&langue=2>. It is possible that in this lecture Deleuze made Kant a Nietzschean of sorts; see Hugh Tomlinson and Barbara Habberjam, translator's introduction to Gilles Deleuze in *Kant's Critical Philosophy* (Minneapolis: U of Minnesota P, 1984). This lecture is close to how I myself came to read Kant by way of Joan Copjec's reading of him.

39 See Copjec, present issue. As a corollary, with Kant, perception is no more a subject–object encounter. The subject is never "in the present" with the object. Rather, every perception now has a temporal gap – an "anticipation of perception" (Deleuze, lecture on Kant).

40 Viveiros de Castro, "Perspectival Anthropology and the Method of Controlled Equivocation" 4.

41 Ibid. 14.

42 See Claude Lévi-Strauss, *The Elementary Structures of Kinship* (Boston: Beacon, 1969) 31.

43 See Alan Schrift, ed., *The Logic of the Gift: Toward an Ethic of Generosity* (New York: Routledge, 1997).

44 Deleuze and Guattari cited in Eduardo Viveiros de Castro, "Intensive Filiation and Demonic Alliance" (undated unpublished manuscript available <http://amazone.wikia.com/wiki/Intensive_filiation_and_demonic_alliance_-_E._Viveiros_de_Castro>) 19. This piece was later published in *Deleuzian Intersections: Science, Technology, Anthropology* (New York: Berghahn, 2010). The two versions are slightly different, so I cite both with appropriate citations.

45 Viveiros de Castro, "Intensive Filiation and Demonic Alliance" (2010) 258.

46 Idem, "Intensive Filiation and Demonic Alliance" (n.d.) 24.

47 Ibid.

48 Cf. Joan Copjec, *Read my Desire: Lacan against the Historicists* (Cambridge, MA: MIT P, 1994) 60.

49 Idem, "*Gai Savoir Sera*" 123.

50 Viveiros de Castro, "Intensive Filiation and Demonic Alliance" (2010) 236.

51 Gilles Deleuze, "The Fold," trans. Jonathan Strauss, special issue on Baroque Topographies, ed. Timoth Hampton, *Yale French Studies* 80 (1991): 227–47.

52 Deleuze, *Logic of Sense* 41.

53 Lévi-Strauss, *Elementary Structures of Kinship* 12.

54 Kojin Karatani cited in Kordela, present issue.

55 Deleuze and Guattari, *Anti-Oedipus* chapter 3.

56 Ibid. 162.

57 See also Lévi-Strauss, *Elementary Structures of Kinship* 13–31, 185–88, 309.

58 Deleuze and Guattari, *Anti-Oedipus* 162.

59 Myth plays the same role as "dishonoring" in their interpretation, which I am not going to delve into here.

60 See Jacques Derrida, *Of Grammatology*, trans. Gayatri Spivak (Baltimore: Johns Hopkins UP, 1976) 103–04. By this formulation, I have in mind another famous case of limit concept: that of taboo in Sigmund Freud's *Totem and Taboo*, trans. A.A. Brill (Amherst, NY: Prometheus, 2000), first published in 1919. Freud's question is how "primitive" societies maintain order in the absence of the written code of law (32) – i.e., the origin of the law. The answer is: those who violate taboo themselves become tabooed objects. That way, taboo acquires a self-warranting objectal force, whose authority does not need to conjure pre-existing codes. In the "disfigured" body of the guilty, taboo makes an objectal "representative" of itself. What had been *excepted* from taboo (for having violated it) becomes an *exemplary* "representative" of the same taboo. A regular example chosen from the "inside" would not do. For it to carry the weight of the entirety of taboo in the body – to embody the very force of prohibition's decisionism qua "differentiator" – the example must be always-already an exception. It is precisely the *undecidability* between exception and example, outside and inside, which turns merely one guilty person to miraculously embody the entirety of the socius.

61 Deleuze, *Difference and Repetition* 143.

62 Bruce Fink, *The Lacanian Subject* (Princeton: Princeton UP, 1995) 77; Deleuze, *Difference and Repetition* 65.

63 By this expression, I am re-invoking the idea of "*a priori past*" or "pure past"; Deleuze, *Difference and Repetition* 81–82. It is not a past that had "existed" (and now manifests itself) but one that only "insists" (82). For Deleuze, *pace* Bergson, without positing this transcendental horizon of a past that had never been a lived-present, we cannot conceive the passage of time at all: "No present would ever pass were it not past 'at the same time' as it is present; no past would ever be constituted unless it were first constituted 'at the same time' as it was present . . . It gives the reason for the passing of the present" (81). In short, some dimension has to be a past always-already. Similarly, in Strathern, one can argue that this a priori dimension named androgyny – precisely as it is "insistent" and not "existent" – is what is "causing" the kind of passage of time she calls "recursivity," and the performative act that subsists in it.

64 Hence, both androgyny and act qualify as the supernumerary One for Strathern. She does not explicitly characterize the former as such, but she does with the latter. "Each act is 'One' act," she writes (II 274). We can infer her androgyny's Oneness from her description of act. The Oneness of act, for her, is emphasized against what I called at the outset performativism. Performativism argues for the contingency of any act, whose outcome reserves plural interpretive possibilities. For Strathern, in contrast, an act is a construction-and-given, cultural-and-natural, and contingent-and-necessary. She writes: "An agent's social position is a multiple one . . . but his or her consequent acts are singular" (II 274).

65 For this reason, I think the proposition such as the following misleading: woman is "infinite to man, not in itself." Richard Reinhardt, "Toward a Political Theology of the Neighbor" in *The Neighbor: Three Inquiries in Political Theology*, by Slavoj Žižek, Eric L. Santner, and Kenneth Reinhardt (Chicago: U of Chicago P, 2005) 61.

66 Deleuze, *Logic of Sense* 50.

67 Ibid.

68 Ibid. *passim*.

69 Ibid. 49.

70 Hence, "all at once" is not to be confused with "singularity," as is often the case. This "all at once" is close to my expression per Strathern: "all particles are already there." See Deleuze's discussion of Lévi-Strauss's concepts of "signifier" and "signified" as series; Deleuze, *Logic of Sense* 48–51.

71 Ibid. 15–16.

72 Contra Kordela, *$urplus*, especially 93–106.

73 Badiou, *Being and Event* 147.

74 Deleuze, *Difference and Repetition* 222.

75 There are moments when Copjec sounds as if she is arguing for this. *Read my Desire* chapter 8.

76 Deleuze, *Logic of Sense* 48–51.

77 As can be seen from my lack of detailed and direct engagement with Kant's antinomies, I suggest this angle in the spirit of provocation or even experimentation, not criticism. My somewhat cavalier sauntering here is entirely thanks to – or, perhaps, at the expense of – disciplined

readers of Kant/Lacan, such as Copjec and Kordela.

78 Copjec, *Read my Desire* 220–35.

79 Kordela, *$urplus* 101.

80 Fink, *Lacanian Subject* 125.

81 Badiou, *Being and Event* 146–48.

82 Wagner, "Afterword" 227.

83 Kordela, *$urplus* 101–02.

84 See Kordela, present issue.

85 Ibid.

86 Wagner, *Anthropology of the Subject* 227; emphasis added.

87 Ibid. 162.

88 Peter Hallward, *Badiou: The Subject of Truth* (Minneapolis: U of Minnesota P, 2003) 68.

89 Kordela would disapprove of this equating, this equation that aligns Strathern-made-Kantian with Cantor's "object"; see Kordela, *$urplus* 103. For her, only in Cantor's set theory can we conceive of an object that is worthy of the Lacanian name "ex-timate." Kant's antinomies (especially the dynamical) do not qualify.

90 Kordela, *$urplus* 102; this undecidability *is* Cantor's definition of infinity. His idea is about how even the "underdetermined" can become determinable when put in relation to each other. In this determination of the underdetermined, the difference at issue is not a structural one but "difference in kind" – the "cut" between determinablenesss and underterminableness; see Deleuze, *Difference and Repetition* 172.

91 I put Cantor's paradox in two propositions in order to re-emphasize the fact that in both Kant's antinomies and Cantor you need both "series" – of the impossible and the prohibited. For Cantor's thesis, in my view, is not just a simple negation of the thesis that infinity is unreachable. He is not therefore saying that infinity *is* conceivable, and that the paradox of inclusion/exclusion or all/not-all is precisely that. He is not saying what infinity *is* or *is not*. Rather, the inclusion/exclusion paradox is merely pointing out that we are "prohibited" from saying what infinity *is*/*is not*. For that, you need both propositions. I believe this is just a slight re-reading of Joan Copjec's detailed clarification of Kant's antinomies. Copjec, *Read my Desire* chapter 8.

92 Modern mathematics, according to Deleuze, is characterized by the function of the "cut" or "limit" (*Difference and Repetition* 172). The cut announces itself by the "universal in number." And "the universal of number as consisting in the cut" constitutes the "next genus of number," the "pure element of quantifiability."

93 This is maintained through progeny because in procreating his son the man is viewed as procreating his own father. This all means that "the corporeality of man is also the physical realm of the cosmos" (II 77). Most impressively, the author is able to demonstrate convincingly how this cosmogony is faithfully reproduced in the practicalities of kinship. Regrettably, it is beyond the scope of the present paper to share this elegant demonstration with the reader. Put simply, through marriage preference man indeed effectively turns himself into a self-birthing being. For "in procreating his son the man . . . procreate(s) his own father," he writes (II 88). All this because of the elegantly simple axiom that one marries one's "classificatory" father's mother – which is, in fact, father's mother's brother's sister's daughter, or, what is the same, FMBSD (87). This results in ego's son marrying ego's (classificatory) mother, and as such practically giving birth to ego himself (II 88).

94 Clifford Geertz, *The Interpretation of Cultures* (New York: Basic, 1973).

95 "In the Iqwaye counting system, infinity is implicit only by interpretation" (II 122).

96 Eduardo Viveiros de Castro and Marcio Goldman, "Slow Motions: Comments on a Few Texts by Marilyn Strathern," trans. Ashley Lebner, *Cambridge Anthropology* 28.3 (2008/2009): 23–43.

97 Deleuze and Guattari, *Anti-Oedipus* 162.

98 Martin Heidegger, "Letter on Humanism" in *Basic Writings: Martin Heidegger*, trans. Frank Capuzzi and J. Glenn Grays (San Francisco: Harper, 1977) 204.

99 Deleuze, *Difference and Repetition*; Lacan, *Encore*.

100 Hallward, "You Can't Have it Both Ways" 41.

101 Copjec, *Read my Desire* 201.

102 Hallward, "You Can't Have it Both Ways" 39–40. Hallward's choice of the password in order to demonstrate the "interpersonality" in Lacan's idea of speech is surprising. The password is the paradigmatic example of Lacan's idea of (in French) *signifiance*, or "signifierness" in Bruce Fink's translation. Fink annotates:

> When Lacan uses the term, it is to emphasize the nonsensical nature of the signifier, the very existence of signifiers apart from any possible meaning or signification they might have . . . [It refers to] "the fact of having effects other than meaning effects." We should hear [from this word that] the signifier defies the role allotted to it, refusing to be altogether relegated to the task of signification. (Lacan, *Encore* 18–19)

In other words, the "coincidence of signification and object."

> The most stimulating expression of this idea is in Derrida's commentary on Paul Celan's work on the theme of shibboleth qua password. In this exposition, Derrida portrays the "machinic" aspects of sovereign power with the password of shibboleth working as its instrument. (Jacques Derrida, *Sovereignties in Question: The Poetics of Paul Celan*, trans. Thomas Dutoit and Outi Pasanen (New York: Fordham UP, 2005) *passim*)

103 Hallward, "You Can't Have it Both Ways" 43–46.

104 Whether Hallward is in fact Deleuzian or Lacanian is not my concern here.

105 Deleuze, *Proust and Signs*; idem, *Difference and Repetition* chapter 2.

106 Deleuze, *Difference and Repetition* 80.

107 Ibid. 82.

108 Ibid. 81.

109 See also Deleuze and Guattari, *Anti-Oedipus* 142.

110 Deleuze, *Difference and Repetition* 79; emphasis added.

111 Claude Lévi-Strauss, *Introduction to the Work of Marcel Mauss*, trans. Felicity Baker (New York: Routledge, 1989) 60–64.

When yin and yang go wrong,
Heaven and Earth are hugely disturbed.
Then comes the crash of thunder,
and fire from the midst of the rains
which destroys the great trees.
Gaining and losing,
the people are caught between them both
and there is no way out.
. . . Gain and loss rub against each other
and start fires beyond number
that burn up the balances of the heart in most people.
The moon cannot contain such fires.
All is destroyed,
the quest for the Tao ends.
The Book of Chuang Tzu

arun saldanha

AGAINST YIN-YANG

the dao of feminist universalism

At a Hegel conference Luce Irigaray once said: "To keep faith with life the universal must manifest and maintain the becoming of living things as they are: *sexed.*"[1] Even if it is easy to see how universalism and vitalism can entwine, the universal is at first sight at odds with the assertion of sex. But when Irigaray says sexual difference is eternal and everywhere, she simultaneously says it has until now been everywhere stifled. Her thought has a rashly utopian dimension which makes her uneasy with the term feminism. With statements such as "I love *to* you" and "democracy begins *between* two" Irigaray means that grasping the irreducibility of sexual difference is also traversing it with love and respect.[2] Not only women have to learn what Irigaray calls in Daoist vein "the way of love," the virtue and pleasure of relating to another without taking, imprisoning, or commodifying.[3] Such reciprocity established at the level of the amorous relationship will extend to the law, the European Union, religion, and the planet.

Understood in its critical aspect Irigaray's project carries on the subversion of the self-transparent subject of the European Enlightenment, of rectilinear history as authored by what Deleuze and Guattari call White Man.[4] To bolster her critique Irigaray develops a naturalist inclination by grounding sexual difference not only in Lacanian psychoanalysis and linguistics but in pagan and Asian philosophies of life. In turning the question of sexual difference *as such* into an endless philosophical project, Irigaray offers a generic political template. Naomi Schor has observed that Irigaray stays committed like Beauvoir and francophone intellectuals in general to the basic tenets of the French Revolution.[5] The present essay likewise argues that universalism is imperative to deal with the catastrophes presently unleashed by global capitalism. While women and transsexuals have more to gain from transforming patriarchy, men would do better too with fewer of the anxieties and less of the competitiveness underpinning their sexuate position.

Two conceptual difficulties arise here: one, how does the extension from love to politics exactly unfold; and two, through which philosophical idiom could sexual difference be universally acknowledged qua potentiality instead of inevitability. Minority women in anglophone countries and women from formerly colonized countries led the way in exposing the striking but unacknowledged biases of white middle-class feminism's claim to universality. Wary of every universalism's intractable partisanship, the feminist and antiracist impetus of poststructuralism has come from a clear departure from Hegel's normative conceptions of essence and the universal.

Yet I want to claim that the feminist reinvention of universalism is necessary for the very reason that universality is – courtesy of climate change, financial mayhem, famine, refugees, and other disasters of capitalism – *more than ever* at the heart of the politico-theoretical agenda. When there are explicit efforts at reformulating universality they are usually (and constitutively) blind to the potentiality of sexual difference. In the work of Alain Badiou, the most cogent contemporary theorist of universality, politics is separated neatly from love.[6] I will argue that universalism *requires* a mooring in sex, love, and other corporeal-psychical differentiations which necessitate a philosophy of life absent in Badiou.

The poststructuralist critique of universalism has a notion of positionality damaging for politics. If what can be politically imagined is only enabled (and thwarted) by a subject's multiple positions in "society," if particularism or strategic essentialism is the best that can be agreed upon, the question is still who within each identity is going to draw up its agenda. The fragmentation towards ever-more-particular identities and attendant political correctness not only risks fascism in Félix Guattari's generic sense but itself essentializes the universal level as beyond intervention.[7] Politics reaches outside one's supposed place(s), outside the fetishism of labels and the spontaneity of hybridization, or it isn't politics but lifestyle. And the more entrenched in the nurturing of particularities, the further politics is from transforming what Guattari recognizes is the ultimate engine of the desire for identity: capitalism.[8]

Universalism does not equal the will to dominate and to silence. Moreover, partisanship should be understood as *necessary* for the practice of universalization, as Marx realized well. Badiou asserts in *The Communist Hypothesis* the wager of politics has for millennia been that domination will be assailed by a universality it cannot contain.[9] While hegemonic universalisms have to obfuscate their particular origins and inconsistency, the particularity of communist universalism is explicit: *workers* of the world unite. Its unfinished, ambiguous, yet potentially sweeping nature makes Marx's counter-universalism forever relevant, while the term "worker" is of course open for redefinition. What poststructuralist critiques misses is that conceived as a process without guarantee and predetermined subject, universalization should in fact be close to their heart.

A similar explicitly partisan and immanent universalism is what Irigaray advocates in relation to the ineluctability of a sexually specific viewpoint. As she writes in a telling passage in *I Love to You*:

> Sexual difference is an immediate natural given and it is a real and irreducible component of the universal. The whole of human kind is composed of women and men and of nothing else. The problem of race is, in fact, a secondary problem – except from a geographical point of view? – which means we cannot see the wood for the trees, and the same goes for other cultural diversities – religious, economic and political ones.
>
> Sexual difference probably represents the most universal question we can address. Our era is faced with the task of dealing with this issue, because, across the whole world, there are, there are only, men and women.[10]

I return to the "problem of race" and the twoness of sex below. Here I will note that Irigaray's concept of difference appears uncertain exactly around the question (marked by the question mark and "probably") whether geography and race can ever trump the importance of sex. This is where she is to be pushed further. Her advocacy for sexual difference has gone hand in hand with an acceptance of the European nation-state as harbinger and guardian of inalienable rights without attending to the violence that the state-form is itself based upon. Furthermore, her adherence to mainstream European notions of Eastern spirituality diminishes the efficacy of both her feminism and her universalism. Guattari's experience in psychiatry showed him that racial tropes enter into everyone's desire. To put it bluntly, Irigaray's whiteness remains as opaque to herself as she says sexual difference is to male theorists.[11]

As Penelope Deutscher has argued, in calling for multicultural dialogue Irigaray neglects to interrogate the boundaries supposed to exist between cultures.[12] We can go further: because democracy and the conferral of rights inherently operate along patriarchal lines they cannot constitute the horizon of politics. Given Irigaray's insistence on becoming we should heed additional caution where her naturalism turns into a gynocentric cosmology in which woman is the immediate wellspring for social transformation. Her move towards a positive imagination of future harmony in twoness is a definitive move away from Freud's pairing of civilization with discontent and aggression. Like the popular conception of yin-yang (twoness is the motor and latency of oneness), Irigaray's spiritual notions of love and community repeat a certain metaphysics of closure she herself criticizes.

Margaret Whitford pinpoints what *I Love to You* means for the Irigarayan trajectory: "What has been described as the 'philosophical terrorism' of *Speculum* has given way to an apparently more law-abiding concern with citizenship and rights, in which the central political concept is love between the sexes."[13] What I want to retain from *Speculum of the Other Woman* is, indeed, its philosophical terrorism, its visionary courting of death. The destructive pregnancy of sexual difference inheres in the fact that man's self-aggrandizement has always belied his proper helplessness. The futurity of sexual difference consists in forcing loose woman's non-wholeness from its concealment beneath man's false universality, and thereby reinventing a truer universality for both. Far from being locked in place by millennia of physical and epistemic violence, woman has evolved – precisely because of that violence – into the quintessential site of revolutionary propensity.

This propensity unfolds in material practice: woman is not a transcendent essence. Universalizability emerges, therefore, from an ontology of immanence. Though Deleuze distrusts universalism as much as he does democracy, his and Guattari's championing of absolute immanence requires subtracting concepts and affects from territory and tradition (Europe, Christianity, Enlightenment), a resistance against what exists not dissimilar from Badiou's truth procedures. If Deleuze's philosophy does not secure a pedestal for communist mobilization, it offers an incubator instead, a space for first *imagining* just futures:

> The artist or the philosopher is quite incapable of creating a people, each can only summon it with all his strength. A people can only be created in abominable sufferings, and it cannot be concerned any more with art or philosophy. But books of philosophy and works of art also contain their sum of unimaginable sufferings that forewarn of the advent of a people. They have resistance in common – their resistance to death, to servitude, to the intolerable, to shame, and to the present.[14]

To remain immanent, universalism requires what Alberto Toscano calls *non-dogmatic anticipation*, a realistic and rigorous attention to how singular struggles gain momentum.[15]

Master signifiers enforced by capitalism – Property-Law-Freedom, God-Nation-Man-Race-Family, Trade-Growth-Technology, Nature bountiless – have pseudo-universal appeal by promising ubiquitous access to some Equilibrium. Many communists and anarchists have yet to fully rid themselves from these promises of transcendence. Their philosophical predecessor, German Idealism, tends to posit history as a self-realizing march of already constituted peoples and soils – civil societies, nations, states, Europe – progressively transcending geographical contingency. Slavoj Žižek offers a revolutionary universalism by cleverly redialectizing this very German teleology, but almost makes retroactive determination of revolution's own conditions of possibility more important than its efficacy as interruption.[16]

In contrast, Deleuze and Guattari attempt to leave teleology aside. Discourse capable of undercutting capitalism proceeds immanently by investigating how master signifiers concretely structure economic, racial, and sexual differentials. Mao said: "You must investigate! You must not talk nonsense!"[17] And the minimum that postcolonial feminism contributes to communism – it contributes infinitely more – is an initial skepticism towards lofty beyonds.

universality as subversion

In a concise essay of the early 1990s, "Sexual Difference and the Problem of Essentialism," Elizabeth Grosz clarifies the convoluted debate signaling the end of second-wave feminism about feminine essence.[18] She remarks how essentialism is often entangled with universalism, biologism, and naturalism, but each of these is nonetheless a distinct discursive strategy through which femininity has been elided, suppressed, and homogenized in the West. Feminists have to be clear on these strategies in order to launch their critique of patriarchy. Whereas liberal-egalitarian feminism is limited insofar as it does not question the white origins of the universalism fueling its struggles for equality, the radical-feminist deployment of essentialism (often based on anatomy and Romantic ideas of nature) faces the challenge of how to conceive existing heterogeneity and open futurity if the characteristics of woman are fixed.

Grosz aligns herself with "feminists of difference" like Irigaray and Gayatri Spivak who conceive of sexual difference as a concept *itself sexually specific,* understanding simultaneously the profundity of patriarchy's reach into the language that feminism cannot but adopt *and* that a working definition of "women" is inevitable. This makes third-wave feminist politics situational, impure, a continuous disentangling (out) from patriarchy through the constant evaluation of positionality. Feminists of difference are defined by a commitment to *subversion* (etymologically, a turning-from-below) which cannot tolerate the sterility of what Irigaray calls the Same.[19] The question now is whether there can be a different kind of sameness, not despite but through difference.

Universalization is hegemonically thought of as the penultimate project of a subject already supposed to exist, a righteous and purposeful agent of change. Change is presumed synonymous with sheer growth in number – us citizens, Christians, reasoners, earthlings, etc. – and the structure of the terrain of social relations (including those of the household) is largely left untouched. Such versions of universalism are premised on the masculine fantasy of a unary subject position, best symbolized in capitalism's hegemonic language, English, as "I," a lone capital letter, an erectitude already vanishing yet full of authority.

Delivered at an earlier Hegel conference, Lacan's famous paper "The Subversion of the Subject and the Dialectic of Desire" was an antiphilosophical provocation taking stabs at positivist psychology, cybernetic theory, hypnosis, ethology, Buddhism, and psychedelic research, all proximate to psychoanalysis but regressing into pre-Freudian ideologies of identity.[20] For Lacan, psychoanalytical science adopts the dialectical methodology from Hegel but – as Žižek shows, precisely by following the dialectic through – it cannot accept the *über*-philosopher's projection of a final coming-to-itself of consciousness. Subverting the Subject – who/which is first and foremost embodied in the false universalism of Man – consists in diagramming the constitutive *mis*-recognition of self and other and the dark desires which consciousness has to repress.

Lacan names the castration complex "the mainspring of the very subversion [he is] trying to articulate by means of its dialectic."[21] All law is ultimately based on the anxiety of having lost and further losing the phallus. Phallus is the fact that the penis is from the beginning a signifying organ and as such an impossible presence like the square root of minus one.

Irigaray will extend this opaque relation between morphology and sexuate position to women but render it positive and direct. What I retain from Lacan's paper is rather the impure reconstruction of the universal subject from and against Hegel. For though he may start out from the concreteness of clinical practice, Lacan is attempting to convince philosophers, hence entering the domain of universality. The subversion of subjectivity is invented through an extrapolation from psychoanalysis to philosophy. True universalism requires the continuous subversion of the substance and the self-conception of the Subject, that is, of sexual-racial and economic injustice.

Ernesto Laclau is one among many who exposes the self-defeating logic of multiculturalism. Far from an ideological illusion, universalisability is to Laclau objectively coterminous with heterogeneous society. Hegel rewrote Plato's doctrine problem of universals by understanding universals not as eternal Ideas but as dialectically extricating themselves as moments within/through particulars and individuals.[22] Laclau's *Emancipation(s)* argues that the universal is not a monolithic realm to be (re)integrated with the particular as it tends to be for Hegel, but nothing more nor less than what every particular lacks insofar as it is constituted by antagonism and negotiation.[23] There can only be struggle – demystification, dissension, coalition – if particular groups claim their ideas and practices have wide validity and challenge those holding power to redefine what is valid. There would be no progress at all if the universal were not a virtual double of every particular struggle for recognition. Now, this unfinishedness of identity is a fortiori instigated by woman, since she is part of (or haunts) every human group, raising the question why we should consider woman an identity at all. My conclusion will ask whether she should instead not be a name for the incompleteness of universality itself.

Universality is not a given sameness-totality of humanity to be (re)discovered, but the impossibility of any social formation or humanity at large to represent "itself" as self-same and bounded. The universal is an empty placeholder, a floating signifier through which particularities assert themselves hegemonically. Ambivalent and contingent, universality exists only in the fleeting collective attempts to determine which people are to be emancipated and how, a representation which will always fail but thereby remain open to contestation. Emancipatory movements propound their universality when they discover their interlinked oppression while building coalitions, like feminists with ethnic minorities.[24]

As the title of Linda Zirelli's essay on Laclau's book suggests – "This Universalism Which is Not One" – an unexpected rapprochement between Laclau and Irigaray becomes possible.[25] But Laclau himself misses what I take to be the fundamental point of feminism. He writes:

> When Mary Wollstonecraft, in the wake of the French Revolution, defended the rights of women, she did not present the exclusion of women from the declarations of the rights of man and citizen as a proof that the latter are intrinsically male rights, but tried, on the contrary, to deepen the democratic revolution by showing the incoherence of establishing universal rights which were restricted to particular sectors of the population.[26]

The semantic ambiguity of Laclau's formulation (for clearly the rights of "man" are indeed "intrinsically male") hints at the fundamental reason why *feminism is not a particularism*: it aims not at emancipating "particular sectors" but the species as infinite series, as more-than-man. Since every social formation is sexually split and reproduces sexually, the welfare of women is a concern for everyone. Whether bourgeois, radical, or poststructuralist, feminism is always-already a relay of universalism. Joan Scott argues:

> To the extent that it acted for "women," feminism produced the "sexual difference" it sought to eliminate. This paradox – the need both to accept *and* refuse "sexual difference" as a condition of inclusion in the universal – was the constitutive condition of feminism as a political movement throughout its long history.[27]

The converse is even truer: universality is only revolutionary when the tension between its own abstraction and the concrete sexual and racial bodies it is actualized in is precisely what keeps it going.

irigaray/badiou

Badiou's universalism goes a few steps further than Laclau's. Like in Laclau, universalization does not work towards wholeness or peace, but Badiou's Maoism prescribes attacking the quasi-automatic recourse to democracy and pluralism. Badiou argues strongly in his *Ethics* for a politics of "the Same." All humans can understand that existing injustice and suffering are perfectly unnecessary.[28] They become stronger by teaming up with disenfranchised others, forming bonds not sanctioned or even thinkable by those in power under banners such as "workers" or "make the 1% pay." True politics goes further than the ethics of corporeal encounter and representation, to find the means for gaining momentum *across* territories and environments. Such internal cohesion of the movement of universalization derives from the scission it effects not just from the status quo but also from the already-coopted reformist forces it mingles with. Mao's adage "one divides into two" sums up the Badiousian version of materialist dialectics.[29]

In a nutshell, universalization has to be *disciplined* in an old-fashioned sense. Bolstered by the mass uprisings across the world in 2011 but understanding why they fail, discussion has finally returned to what better "worlds" can be hoped and strived for. Irigaray's pursuit of sexual reciprocity through "peaceful revolution" now appears somewhat reactionary. In his introduction to *Ethics* Peter Hallward singles her out to show that the consensus on the sanctity of difference merely extends the individualizing power of the capitalist state:

> Whereas Badiou's subject, through fidelity to an event of love, seeks to subtract a truth of sexual difference from all positive or culturally validated factors (i.e. from what can be known of sexual difference), Irigaray's goal is precisely "social and cultural sexualization for the two sexes". This effort leads her to assert "women's right to their own specific culture", complete with a specifically feminine "sort of social organization . . . a religion, a language, and either a currency of their own or a non-market economy".[30]

It is incontrovertible that Irigaray's onto-politics lies squarely within what Badiou's *Logics of Worlds* calls democratic materialism for which "there are only bodies and languages." But her radical-feminist line is more fruitfully understood as analysis of patriarchal-capitalist epistemes rather than prescription for change. It is often forgotten that the sting in *Speculum*'s critique of Freud – the sting which got Irigaray excommunicated from the Lacanians and from Vincennes – drew not from Beauvoir but Marx and Engels. Throughout her work Irigaray acknowledges the deep entwinement of patriarchy with ownership, commodification, inheritance, exploitation, resource extraction, war, and the split between public and private.[31] In contrast, Badiou offers almost no analysis of the socioeconomic conditions compelling him towards communism.

The risk in Irigaray of colluding with conservative moralism or a sentimental ecofeminism is real. But her work can also be understood to illuminate dimensions of sexual-political economy which Marxism and psychoanalysis have not been able to illuminate. In interactions with the Italian Communist Party, Irigaray has addressed the evils of capitalism as culminations of many centuries of patriarchy. These interactions themselves actualize what Irigaray understands under sexuate love and democracy.[32] It is hardly worth mentioning that the aptly titled collection *Lenin Reloaded* says nothing about gender, nor that the seventeen chapters are by men. Nor is it surprising that a book called *Lacan and the Political* does not consider feminism political.[33] Irigarayan analysis reminds revolutionaries that intervention is to be grounded in investigations laying bare very precisely what is wrong with the situation from which political truth will extract itself. This includes sexism in the (revolutionary) left itself.

The construction of political subjectivity should not have to await a mystical event of conversion. This is what Badiou effectively commends in his book on the Pauline "foundation" of universalism.[34] In sharply distinguishing the Event precipitating true universal becoming from the mere Being of the status quo, he gives us a tremendously powerful notion of belief. But the price he pays for a deliberate neglect of empirics is that there is no cognitive mapping available – to use Fredric Jameson's Althusserian-geographical term – of and for universalization to actually *take place*.[35] Mao sees better than Badiou that theory anticipates practice through study.

In exposing the vastness of sexual difference Irigaray *already* engages the universal, not as a sterile oneness but as a reservoir calling for further intervention. If femininity is understood traditionally in psychoanalysis and sexology as the mysterious, the leaking, the incomplete, this makes it/her the starting point for radically transforming humanity "as a whole" – except that it is to the precise extent that humanity is *suspended as* whole, as eternally given, that such transformation can happen. We should nevertheless insist on the term "feminism" not only because women have more to gain from disassembling patriarchal power but because it is the feminine subject position that is lucid enough to guide such disassembling. Men have only shown their stupidity in entrusting themselves the business of formulating universality.[36]

What Irigaray's de/reconstructive evasion of Hegelian teleology amounts to is that woman is the *only* particularity which is an inexorable component of any subversive project. Despite its name and history feminism is not merely a struggle for the emancipation of an identity but the struggle for generic emancipation as the *immediate result* of dismantling the subordination and vulnerability to abuse amongst half of humanity. This means that though universalism greatly exceeds feminist politics (it also, and more urgently, subverts capitalism and racism), universalism cannot but be feminist, and feminism cannot but be universalist. It is imperative to step into the arena where eternal ideas are fought for and link sex/sexuality inexorably with the aims of Kant, Marx, and Sartre. As a result feminism is never *simply* cosmopolitan, revolutionary, or existentialist. Universalist legacies are to be adopted thuswise by stripping the auras of infallibility often conferred upon intellectual father figures.

Improvements of the human condition come about when agents of change constitute themselves from scratch "every time" an unjust situation calls for it. These "times" engender epistemic and political revolutions, or waves, as in feminism. On the one hand universalism has mostly been gender-blind. On the other, the Eurocentrism and imperialism of white liberal feminism continues almost unabated to shape international development policies while in the West women's and gay rights increasingly go together with nationalism, Islamophobia, and consumerism.[37] Feminists and queers have understandably shunned pretensions to universality, but they risk leaving the arena open for the likes of the Dutch-Somali neoconservative Ayaan Hirsi Ali.

At this juncture feminism faces three historical givens: first, the politics of difference has been commodified by a neoliberalized multiculturalism; second, this politics does not prevent corporate capital from defining the shape of globalization with a cynicism not seen since the nineteenth century; third, a geopolitical shift "in favor" of East and South Asia is going to affect the very definitions of humanity, democracy, faith, and sexual difference. Every wave of feminism constitutes itself by examining both how patriarchy has changed (remembering it operates differently in different cultures and social strata) and the impasses of previous waves. Feminist universalism enters into the struggle to enact what could and should be true, just, beautiful, pleasurable for everyone, *and* to distinguish how male, female, and intersexual bodies are vehicles for universalization. It is both empirically obvious and theoretically central that sexual particularity in no way rules out a universal capacity to transduce the universal.[38]

In short, we have to ready ourselves for *fourth-wave feminism*. First there was the equality-feminism of Wollstonecraft; then the identity-feminism exemplified by Simone de Beauvoir's existential politics and lesbianism; third, the difference-feminism advanced by Grosz, Spivak, Rosi Braidotti, etc. Now I am hopeful for universal-feminism.[39] Unfortunately Irigaray drifted back towards the first two waves after triggering a third- and even fourth-wave tsunami. Her original radical demand – "impossible yet necessary" as said by Lacanians – was a rethinking of the entire framework through which love and justice are distributed. Nothing is more subversive than that. If she herself directs her disruption back into the folds of liberal democracy that does not have to be its only destination.

the mysterious cleft, nondoing, state desire

We now need to develop universal-feminism in greater conceptual depth. Accelerating ecological catastrophe has an intrinsically unjust impact on populations, with poor brown and black women and children most vulnerable. This makes an embedding of human differences within the biophysical planet indispensable. In *Between East and West* Irigaray seeks to break courageously but humbly from the mold (container-fungus) of modern European concepts of nature and spirituality. "Familiarity with the tradition of India can put us on or put us back on the way, or, at least, challenge us."[40] She criticizes Schopenhauer for misunderstanding India, but her own interpretations of yoga and Indian philosophy also flirt with orientalist truisms. Misunderstanding, however, is creative. The problem is the wrong choice of Oriental tradition. Twoness and sexual difference are explicit starting points for Chinese not Indian philosophy. The Daoist valorization of feminine qualities and forces is more systematic than the "aboriginal," "pre-Aryan" matriarchal religions that Irigaray supposes preceded Hinduism.

The virtues that Irigaray propounds of self-cultivation, quietude, respect for natural cycles, and love within the couple are all basic to Chinese philosophy. They do not follow from India's pessimistic premise of suffering as caused by desire. Indeed, Irigaray forgets that the Judeo-Christian concepts of guilt and self-sabotage she wants to escape from have Indian precursors. The "between" of her book's title names the always-already of the Indo-Irano-European realm. Irigaray furthermore omits caste, reincarnation, the patriarchal functions of temple and marriage, and the rich mythology of warfare, all inextricably Hindu. Compassion is not straightforward in Hinduism; central to Buddhism it is nonetheless not a political guideline. Aimed at harmony with nature, Irigaray's work closely resembles ancient Chinese political thought and especially the concept of ☯.[41]

Of her turn towards pre-Socratic themes of nature through Nietzsche and Heidegger, Irigaray says:

> I had thought of doing a study of our relations to the elements: water, earth, fire, air. I was anxious to go back to those natural matters that constitute the origin of our bodies, of our life, of our environment, the flesh of our passions. I was obeying a deep, dark, and necessary intuition, dark even when it is shared by other thoughts.[42]

Irigaray and Daoism both center their cosmology in fluidity and obscurity, that is, reality's feminine aspect. European philosophy has strenuously disavowed its linkages with Asian intellectual traditions, and the Daoist metaphysics of darkness contains still largely untapped resources for the subversion of traditional Western essentialism and rationalism.[43] However, the belief in an intuitive return to elementary darkness thwarts subversive universalism. Inner darkness can be intuited, perhaps, but not "obeyed." As immanent negativity the obscure cannot be the object of any intentionality. The infiniteness of the cosmos is not an origin to be inhabited – this would in any case mean psychosis or death – but a primordial excess which has to be presupposed for any affirmation of possibility to occur.[44]

All Chinese philosophy is in sociological terms the translation by the ruling classes of millennia-old animistic and ancestor worship into theory useful for militarized feudalism. The resulting patrilineal biopolitics is most systematically and influentially prescribed by Confucius (Master Kong), as in the following passage of the *Analects*:

> Duke Jing of Qi asked Master Kong about government. Master Kong replied: "Let a ruler be a ruler, a subject a subject, and a son a son." "Excellent!" said the Duke. "Indeed, if a ruler be not a ruler, a subject be not a

> subject, a father be not a father, and a son be not a son, even if there is grain, shall I manage to eat it?"[45]

During almost two thousand years the *Analects* was the official handbook for bureaucracy and civility, testimony to its uniquely conservative capacity. Daoist scholarship emerged in the fourth and third century BCE as a questioning of this Confucian rigidity. The interplay of the two within is often itself considered in terms of yin vs. yang, feeling vs. reason, folklore vs. the state. Both Confucians and Daoists were persecuted under the short but harsh Qin empire (221–206 BCE), which united the land for the first time and from which the name China derives. The famous self-nominated First Emperor adopted the newly systematized Legalist philosophy, which was concerned exclusively with perpetuating imperial power through the severity of law.[46]

Daoism posits a primordial pathway for mortals enigmatically mapped out by nature herself. Dao, the Way, had been a central notion well before Confucius. It expresses the fact that everything flows ceaselessly and in its own manner, self-organizing according to hidden laws. For many scholars this makes water the privileged element in Chinese philosophy.[47] One only finds the Dao through living life's flow. What is sought is not redemption from sin but wisdom and peace. Whereas monotheistic religions presume a sharp historical cut, Daoists walk an already trodden path. There is no roaming, there are no road blocks. While an individual's path has to be found in a forest of elusive possibilities, it is a path with definite direction, boundaries, width, and slope, to which one adapts. The Way is not created from scratch with a machete.

Written down some time in the Warring States era (*c*.475–221 BCE) and ascribed to Laozi, the "old master," the *Daodejing* (Book of the Way and the Immanent Force) is Daoism's central text. Its first lines argue that language is intrinsically alienating: "The Dao that can be told is not the eternal Dao" (Ryden).[48] The basic principle of order existed before there was space, time, matter, or god, and forever makes these possible. Inexhaustible and omnipresent, the Dao has some affinity with the Western concepts of void, except that like chaos it already contains incipient organization. The Dao is a virtual river from which material reality is woven. The river of life, weaving, birthing: all these are obviously correlated as feminine aspects. As chapter 6 of the *Daodejing* says suggestively:

> The gully's spirit does not die
> She is called "the mysterious cleft".
> The gate to the mysterious cleft
> Is called "the root of heaven and earth".
> Weaving on, continuously existing
> Use of her shall never end. (Ryden)[49]

Becoming a sage is to penetrate back into the soil-womb, bypassing the obfuscation of human law. Interestingly, text and textile are closely linked in Chinese thought and dao also means speech. Like in Schelling and the Book of Genesis, in the beginning there was the Word, but it was spoken by the Feminine. The *Daodejing* does not posit a transcendent God who created reality with a lucid command but an "abstruse" semiotics (chapter 14, Ryden), an originary unnamable power of communication constituting reality itself.

Where the Dao pertains to humans it becomes linked to the second central concept of Daoism, *de*, often translated as virtue, but more metaphysically Immanent Power. Also connoting the verb "to attain," *de* is the masculine vitality borne from the feminine nonbeing of the Dao. Virtue does not arrive through ritual or thinking and talking about it but through intuiting the primeval communing of the spheres of man, earth, and cosmos. As a work in ethics the *Daodejing* teaches an unlearning of common sense, ruled as it is by greed and short-sightedness. By living virtuously the Daoist vanguard will spread virtue universally: "If a nation is centered in the Dao, if it nourishes its own people and doesn't meddle in the affairs of others, it will be a light to all nations in the world" (chapter 61, Mitchell).

In Daoist vitalism it is not abundance that is celebrated but void: "We shape clay into a pot, but it is the emptiness inside that holds whatever we want" (chapter 11, Mitchell). It follows that political power is attained by exerting not strength but serenity and patience: "Nothing in the world is as soft and yielding as water. Yet for dissolving the hard and inflexible, nothing can surpass it" (chapter 78, Mitchell). The Buddhist concept of desire was grafted onto Daoist-Confucian conformity. Desire is all over the place. Guided by the Way man empties himself from indulging in women, commodities, and status which, given his position in the social-natural order, he has no "way" of maintaining anyway. Hence the famous Chinese endorsement of withdrawal, effortless action, doing not-doing (*wei wuwei*). Misrepresentations of *wuwei* fuel white Sinophobia (blind obedience; opium dens; evil despots exploiting women, eunuchs, animals, dissidents, or laborers; toxic commodities), but Daoism's "aim" to nurture inner stillness and become liberated from the injunction to choose is fully practical and socially cohesive. The dao of the world is to work well, not to evince predestination but simply to avoid struggling with one's place. The handful of sages – legendary men and a few women who gained immortality through meditation and alchemy – are strictly exceptions, defining the rest of us in our ordinary finitude.

Despite their anarchic and skeptical dimension, Daoist writings are as much textbook examples of what Massumi in his foreword to Deleuze and Guattari's *A Thousand Plateaus* calls "state philosophy" as the Confucian ones.[50] The humor and provocations of Zhuangzi (Chuang Tzu, 370–301 BCE), the first and greatest of Daoist authors I cite in the epigraph, ultimately aims to keep everything in balance. Whether Han, Qin, Nazi, Bushite, or communist, a state feeds on "state desire" directed at the "vanishing body" of the male despot, and I would add ultimately at the identification of race and territory. The state channels the flows of desire amongst the masses into what is useful to it – hydraulics, production, war, hierarchy – and into its patrilineal order. "State desire is by nature patriarchal."[51]

The political paradox of Daoism is that there is in the sage a becoming-imperceptible in the Deleuzian-Guattarian sense *but thereby* the stabilization of tradition. "Those in antiquity who were well-versed in the Way were recondite and abstruse, so deep that they could not be discerned" (chapter 15, Wagner). By imagining a primordial feminine Beyond which one can succumb to Daoist philosophy harmonizes absolutism and socioeconomic inequality. Gynocentric, ecstatic spiritualities fulfill this integrating function across the world today. States are intrinsically suicidal as Deleuze and Guattari say, but their death can be deferred by a vitalist construction of metaphysical femininity.

sexual difference writ large

It is in the concept of yin-yang we find the most proximity between Irigarayan naturalism and Chinese philosophy. Covering everything from colors, animals, organs, and planets to political history, Chinese divination has drawn elementary tables since Paleolithic times, most famously in the *Yijing*, the Book of Change. The five elements (*wu-hsing*) – earth, wood, fire, water, metal – are both agents and phases, entitled to overcome the rule of a previous element in eternal cycles. Han historians designated the Yin Yang School as a separate philosophical movement and stated that Zou Yan (Tsou Yen, *c*.350–240 BCE), an official philosopher of the Qi state, integrated yin-yang thinking with the Five-Elements School. Until Lévi-Strauss there never existed a more encompassing vernacular structuralism. The syntagmatic (correlative) chains of binary oppositions found throughout ancient Chinese literature enmesh cosmology, morality, numerology, and government. Under the Han dynasty binary equivalences came to permeate everything. Across the equivalences, the penetrating principle is always associated with masculinity, the yielding with femininity. They are always complementary.[52]

Yin-yang is steeped in the mores of sedentary agriculture, which Chinese scholars sharply distinguish from the uncivilized nomads. Etymologically *yin* refers to the shaded side of a hill, *yang* the sunlit. Halves of the same hill, they swap places as the sun moves overhead. In animist worship, seasonality and the sexual division of labor conspired to make the pair signify female/male, death/birth, tiger/dragon, etc. With women working inside during the winter months and men on the fields in the summer, *yin* became associated with the inner, the dark, and the turbulent; *yang* is by contrast light, reasonable, productive, the sphere of politics and business. The male principle is not inherently better: order is obtained when negative *yin* and positive *yang* are balanced. By foregrounding the feminine as separate and primordial Daoism nevertheless conserved deeply patriarchal notions of cultivation and order. Daoist sages can be called the "gardeners of the cosmos."[53] Regional indigenous religions and principles from Buddhism became integrated, lending lasting consistency to China's medicine, alchemy, astrology, martial arts, and cinema.

Irigaray's sexual difference correlates fluidity/solidity more than dark/night. *Speculum* argues that the psychoanalytical depiction of woman as the formless, excessive, hysterical sex merely extends patriarchy's abhorrence of menstrual blood and the womb.[54] The essential relationship woman has to rivers and the sea becomes a metaphysical program in *Marine Lover* and "The 'Mechanics' of Fluids."[55] It is crucial that this project of glacial proportions directly involves a hopeful reaching out to men (here Nietzsche):

> In me everything is already flowing, and you flow along too if you only stop minding such unaccustomed motion, and its song. Learn to swim, as once you danced on dry land, for the thaw is much nearer at hand than you think. And what ice could resist your sun? And, before it disappears, perhaps chance will have the ice enflame you, dissolving your hardness, melting your gold.[56]

Irigaray writes sexual difference ultra-large, but how femininity and masculinity map onto natural dualities is not stable. Alison Stone ties Irigaray's poetics to German Idealism to procure a naturalist realism of sexual bipolarity.[57] Confirming the worst fears of Irigaray's critics, Stone argues that sexual difference is not only natural but there are essences to man and woman. Drawing on Schelling, she thinks that bipolarity pervades nature as potentiality and comes to fruition in human sexuation. Schelling's *Naturphilosophie* is arguably where European philosophy comes closest to a "dark" interpretation in yin-yang thinking. His somewhat unfinished speculative realism takes issue with empirical science and Kantian subjectivism, expanding purposefulness towards nature herself. Nature is terrifying, gushing onwardness, an eternal primordial abyss which constrains herself into particular finite beings. Purposefulness includes a violent madness as an ineliminable fringe of all process. Many have pointed out parallels with psychoanalysis, and Schelling is pivotal to Žižek's ontology.

Of interest to Stone is the dualization, not the irrationality of nature. This is not to say that the Kantian thing in-itself is sexed, rather patriarchy has so thoroughly sexed perception that entire dimensions of the world have been forgotten. Sadly, Stone does not engage with Darwin. In knitting sexual difference to evolutionary agency Grosz fills a crucial gap in Irigarayan ontology.[58] Understood as spontaneously self-differentiating, nature does not allow for essence and duality without multiplicity and temporality traversing them. Stone and Grosz provide Irigaray with a monistic concept of difference in which sexual difference is emergent from an all-encompassing background of vital tendencies.

Something similar is found in Daoism. A key statement is chapter 42 of the *Daodejing*:

> The Way generates the One
> The One generates the Two
> The Two generates the Three
> The Three generates the myriad things.
> The myriad things recline on yin and embrace yang
> While vacuous qi holds them in harmony.[59]

To tease out the radical possibilities of these lines let us turn to the interpretation of the prodigious scholar Wang Bi, who died aged twenty-three in 249 CE, not before establishing the "Neo-Daoist" tradition of Dark Learning or Dark Studies (*Xuanxue*) following the doctrine of nothingness in Mahayana Buddhism. On Rudolf Wagner's authoritative translation the One is for Wang Bi the great formless and generative *negativity* to which all beings ("the ten thousand entities," *wan wu*) return.[60]

Being unfolds dialectically. The second line implies that once the One as void is conceived there is something rather than nothing, which also means One doubles into Two (void plus its name). The duality of name and thing is being as such from the perspective of speaking creatures. But then being and nonbeing (the nothing that was) together produce their dialectic, the Three, from which all of physical reality is spun. That is to say, things submit to and are nourished by nonbeing (at this level called *yin*, shadow, or heaven just before dawn), while they develop by seizing being (*yang*, light, earth). The last line seems to say *qi* (life-energy, but rendered "ether of emptiness" in Wagner) is another name for the Three qua generative tension.

What about the first line? Most abstractly, the entire cosmic dynamic is kickstarted and directed by the Dao, which I take to refer here to prenominal pathliness. Even nonbeing has to be preceded by the bare possibility of becoming-ushered-in. And chapter 25, thinking "backwards," presents an even more enigmatic metaphysical level: "The human being takes the Earth as model. Earth takes Heaven as model. Heaven takes the Way as model. The Way takes That-which-is-of-itself-what-it-is as model." Wagner argues that the level of the That-by-which (*ziran*) is reality's ultimate negatability by itself, of which dao is one aspect and imperceptibility (*xuan*) the other.[61] The That-by-which is not God; it is the *immanence* of the dialectical unfolding of the One through its negative essence. It can be glimpsed momentarily by the trained mind.[62]

Three observations are in order. First, there are still largely unexplored commonalities between Greek, German, and Chinese philosophies of nature, which would benefit Irigaray's sexualization of the cosmos. Second, between the third and sixth centuries a dark vitalism was dominant in China for which life was not plenitude but obscurely led hollowing-out.[63] Sexual duality is for Wang Bi derivative of a more fundamental and logical noncoincidence of the Way with itself. Things exist by negating their own negativity. Third, the site of philosophical contention is precisely the place where One is split irrevocably. Žižek senses the ontological stakes:

> This split [of One into Two] has nothing whatsoever to do with the premodern notion that, at all levels of reality, an ontological Whole is always composed of two opposed forces or principles which have to be kept in balance (from Yin and Yang to social freedom and necessity). The Hegelian Twosome, rather, designates a split which cleaves the one from within, not into two parts: the ultimate split is not between two halves, but between Something and Nothing, between the One and the Void of its Place.[64]

Now, no other premodern ontology than Neo-Daoism posits a Whole as immediately divided by its negativity. Furthermore, the *xuanxue* dialectic as tertiary and notionally elusive could potentially be more profound than the Hegelian Twosome. Žižek is correct, however, that Asian philosophies at the practical level aim at keeping forces in balance. For Wang Bi, "the feudal, hierarchical structure of society is naturally ordained and thus should not be tampered with. For the system to work properly only requires the ruler to be as though he were not at all."[65] Conversing about the mysteries of nebulous and unnamable profundity is an elite activity. Under the Han and after, the ontological embracing of radical negativity has been serenely apolitical. If for Schelling, Lacan, and Mao on the one hand, and Daoism on the other, void is constitutive of reality, the former endeavor to pry open the void's non-identity in ever new ways (such *is* modernity), while the latter tries to reconcile being with nothingness by intuiting their equal but different contributions to reality (such is tradition).

This is why Žižek tirelessly delineates true dialectics – the materialism of psychoanalysis and Marxism – from the obscurantist idealism of yin-yang thinking. Early on in *The Parallax View*, for example:

> The key problem is that the basic "law" of dialectical materialism, the struggle of opposites, was colonized/obfuscated by the New Age notion of the polarity of opposites (yin-yang and so on). The first critical move is to replace this topic of the polarity of

> opposites with the concept of the inherent "tension", gap, noncoincidence, of the One with itself.[66]

Whereas the New Age pseudo-dialectic is defined by polarity, the materialist dialectic commences with the paradox of monism: how do you name or see the One-All if you cannot step outside it? Chinese and Indian scholars thought this paradox through much more than Žižek permits, but we need first to understand why sexual-cosmic polarity is obfuscating – why it's so appealing.

sexuation and the chinese dialectic

As Lacan himself has drawn on Chinese linguistics and concepts, especially in the playful Seminar XVIII on semblance, we can risk short-circuiting Daoism and sexuation.[67] The dialectic of desire that is yin-yang commences with the starkest difference possible with no in-between or neutral standpoint possible. The black blob in the inferior position is woman, sub-verting the white swirl of man. Both look like oozing eggs sunny-side-up, or spermatozoa, or tadpoles with large eyes. They are not eyes, however, but the what-is-more-than-itself of each, two kinds of traumatic kernel. The white dot within the woman-blob is *objet petit a,* the unlocatable stand-in for man's original lost object which makes every real woman more-than-woman. The black hole within the man-blob is the symbolic phallus around which woman's desire forever twirls, standing for *her* original lost object. Man is already castrated and woman not-whole. They dance around/in each other constituting a tragic tautological groping from which none can ever escape.

Now, sexuation as thought by Lacan pertains to the human species. If psychoanalysis has its own form of correlative thinking (missiles and dragons are masculine, tunnels and cats feminine, and so on), it is meant to explain psychical processes, not the world of the ten thousand things. While he is interested in what Chinese philosophy (especially Mencius) says about language, Lacan himself strongly refuses the identification of yin-yang and sex, noting that phallic asymmetry prohibits all bipolarity.[68] Male and female jouissance are *both* determined in relation to the phallic function Φ of which a is in no way the counterpart. Yin-yang endorses convergence at the extremities of opposition – man reverses into woman at the height of his power and vice versa – whereas for Lacan man and woman forever bifurcate. There is no polarity in sexual difference, because polarity implies a shared sphere, hence symmetry and reversibility.[69]

Whereas synthesis is the forever-deferred objective of the Chinese dialectic, in Lacan's dialectic the unthinkability of synthesis is what keeps sexual differentiation going. Man, insofar as he is positivity, (ideo)logically assumes prior comprehensive totalization and transparency, while woman, as negativity kept from universality, lacks overarching identity (so should be properly written ~~Woman~~). There are only singular women while men are burdened by preposterous universality. Woman's desire of the phallus leads her through and beyond it.

> The fact remains that if she is excluded by the nature of things, it is precisely in the following respect: being not-whole, she has a supplementary jouissance compared to what the phallic function designates by way of jouissance.
>
> You will notice that I said "supplementary". If I had said "complementary" what a mess we'd be in! We would fall back into the whole.[70]

The task of feminism, as Joan Copjec sees it in a piece commemorating the journal *m/f,* is to follow the bifurcation of woman ever further from man (and Man): "It has always been in the name of harmony (of insuring against woman's being wolf to man) that woman has been conceived as the complement of man, that which completes, fulfills, and heals him."[71]

Irigaray, too, militates against isomorphy between man and woman but for her woman is supplementary in relation to man in a more

Derridean sense.[72] Man has to tear himself from (a) woman (his mother) to be man at all, which makes woman dangerous to his complacency. And yet Irigaray's insistence on the possibility of meaningfully talking to and touching the other does reintroduce a concept of complementarity. Irigaray's universalism is predicated on inventing forms of sociality *across* sexual difference. The way out of alienation for women and men passes through each other, although that requires women first gain the rights and spaces to signify (by) themselves. The "culture of difference," Irigaray conjures, is contingent upon woman supplementing/supplanting/displacing man who has labored judiciously against her coming so close.[73] Still, deconstructing even a misogynous genius as Nietzsche entails some sort of intimacy, revealing a secret desire to be accepted by the other, by man.

Irigaray's attempt is to rewrite 2,500 years of metaphysics and science from women's point of view: no small feat. She never completely loses sight of antagonism, but in moving the emphasis from critique to transcendence she ventures towards an ethics of mutual dependency. Developed especially in *To Be Two* is the notion that the male other is woman's necessary partner in becoming, while inversely man is in greater need of breaking with ho(m)mosociality.[74] Such complementarity fits the legal apparatus of the heterosexual couple presupposed by capitalism and Eurocentric directives on production and reproduction in poorer countries. For keeping a subversive notion of sex I prefer to stick to Irigaray's 1970s radicalism. The tight knot she then invented between feminist epistemology, ontology, and politics is of tremendous and (per definition) under-explored significance to radical thought. All the more reason to trace it carefully.

sexual difference, racial difference

Racism is primarily not psychological but economic. When race is confined to mere prejudice, to drives concealed by the said, it largely escapes critique.[75] Foucault concludes the first volume of his *History of Sexuality* by showing how the management of sex already implies the management of bloodlines. Institutions have to decide who sleeps with whom and whose babies are worthwhile caring for.[76] Power therefore racializes on the genetic level. There are three problems without which a reconstruction of universalism will be weak: race as universalism's repressed real; race as theosophical obsession; and the inevitably racialized corporeal support of universalization.

In any universalism worthy of the name differences the capacity to grasp the universal is considered contingent on learning and situation, not innate. In order to maintain this belief universalism tends to leave unexplained the material origins of its privileged speaking position, while imperial universalisms already presume their neutrality. The inflation of particular accomplishments to universal relevance is knotted, like a balloon is, with the disavowal that other particularities provided labor, resources, and even many ideas towards those accomplishments.

This historico-geographical interconnectivity is systematically denied in white universalism. The hegemonic view that the Industrial Revolution was homemade and turned Northwestern Europeans into pioneers of globalization (not the Chinese, Arabs, Mughals) is not simply chauvinist (ideological) but racist (economic). By obfuscating the generative importance of primitive accumulation from/in the colonies this view secretly explains white ascendancy by innate superiority after all, while legislating further exploitation. Eurocentrism is directly biopower. Master signifiers such as Progress, Growth, Technology are not just fake universalities making capital circulate in the service of the longer, happier lives of lighter-skinned bodies and imply that their provenance makes them *rightfully* universalizing.[77]

Inflationary white universalism is pricked by the materialities of famine and dispossession which it has had to befuddle. Though this inflation is more than complex enough in Kant and Hegel to warrant ongoing study, it becomes ugly in climate change talks, IMF adjustment programs, and the occupations of Afghanistan and Iraq in the name of human and women's rights. White universalism therefore stretches from philosophy to trade policies to the

Abu Ghraib perversities. The more universalism insists it speaks for everyone the more its racist underbelly has to be suppressed. Race has a social *reality* as much as a Lacanian *real*. Intrinsic to and fueling racial signification is a blockage to encountering the mechanisms of capitalist inequality. The multiscalar segregations and fantasies of race are sutured by what they cannot secure – the steadfast categorization of human phenotypes.

Under capitalism sexual difference necessarily doubles into racial difference, though they are irreducibly different modalities of differentiation. Race is the counter-intentionally projected object-cause, on an intercontinental level, of all attempts at cultural belonging through symbolic identification with a body type. Traumas of slavery and genocide accompany *every* nationalism and city, and all universalizing ideologies such as multiculturalism and freedom of speech. If the real of sexual difference is the unrepresentability of desire, the real of racial difference is, in addition, the unrepresentability of capital. Calling racial difference a "purely symbolic" effect of the impasse of sex (the only real real), as Kalpana Seshadri-Crooks does in conventional Lacanian vein, shrouds the ingrainedness of bourgeois sexuality in capitalist colonization.[78]

Unlike hegemonic universalisms, race is of explicit concern to the transcultural eclecticism of theosophy, although after 1945 not usually qua "race." Theosophy is understood here somewhat like Žižek's archenemy "obscurantism," occult subversions of white intellectual modernity from within, sometimes penetrating into exoteric philosophy and theology (hence Schelling's influence on Žižek himself, or Jung's on Deleuze). Theosophy reinscribes the Judeo-Christian theme of the chosen people onto a vision of a future *hybrid* master race. Repeating assertions on the darkness of God, primordial androgyny, the afterlife, etc., theosophies from Jakob Boehme to Timothy Leary and ufology entail a eugenic fantasy of an enlightened section of the species entering evolution's transcendent phase. Notorious in this regard is the astonishingly convoluted Aryanism in *The Secret Doctrine* of Madame Blavatsky, founder of modern theosophy.[79] Instead of being disavowed, racial and sexual difference are constitutive of these subterranean discourses, whose universalism paradoxically turns into fascism insofar as membership is policed by its secret doctrine itself.[80]

Feminist universalism has to deter the seductions of such cosmic biopolitics. When Irigaray is married to Jung in order to reinvent a maternal, Platonist essentialism of alleged a prioris of the feminine, the concrete antagonisms of patriarchy revealed by Irigarayan (as well as Marxian and Foucauldian) analyses tend to be excised.[81] More seriously, there is a danger in the Jungian insistence on a *hierarchal access* to the eternities beneath myths. Suffice to quote a 1934 speech by Joseph Goebbels, with Jung probably in the audience:

> Today I say that *Mein Kampf* has to be called a scientific book... Whoever reads the book and the speeches of the Führer and studies his way of being will recognize that he has something that most of us lack: Jung calls it intuition. It is more important than all science.[82]

But more than simply expose Jung's anti-Semitic leanings and endorsement of Hitler, we have to investigate how racialism ensues logically from old German interests in alchemy and irrationalism, how it blends with psychoanalysis as well as universalism and ecofeminism, and whether communist deployments of mysticism are conceivable.[83]

The third way in which race figures in universalism subtends the first two: political truth is both haunted and recognized by the contingency of its corporeal support. This contamination, examined extensively by third-wave feminism, is not exactly denied by left-wing universalists but insofar as they endorse a Cartesian-like desubstantialization of subjectivity it is considered epiphenomenal. The main concern of Badiou's "phenomenology" or even "physics" in *Logics of Worlds* is how a body comes to support an Idea formally. His body has no senses or shape. Constituting itself through decisions alone, per definition it sheds its animality. In an immanentist framework the phenotypic and geographical specificity of bodies cannot be absolutely incidental to their being swept up by an Idea. What is needed is a theory of how bodies come to *conduct* an Idea, their density and dynamics enabling, resisting, and shaping its propagation. An Idea fashions bodies by mutating itself to accommodate their particularities. Universalism has to be an unstable superposition of Badiou's transcendental scheme upon the historical-materialist investigation of desire and capital.

Rather than brushing aside the critique of universalism's conspicuous whiteness as naively substantialist, it should be possible to bring that whiteness into the fold of the concept of universalism itself, not as dialectically implicated-negated but as the irrevocable structure which was capable, by virtue of its material privilege, of formulating the strongest universalisms ever. Europe's arrogance and avarice instills shame, then opportunity. White Man is the cannibalistic subject of world history, and left-wing universalism cannot deny its own indebtedness to him. It can now contest this Man by passing through the reals of sexual and racial difference which he has had to deny.

In a globalizing world explicitly Eurocentric universalisms will look increasingly provincial. When Žižek decries Mao's "pagan wisdom" and "cosmic perspective," he overlooks that Mao mostly follows a dialectical-materialist philosophy of science not Daoism.[84] Universalism has yet to learn from Chinese and other non-European philosophies. From Confucius we glean honesty, courage, perseverance. Daoism contributes to thinking of social duty as given through a non-anthropocentric future. Even Legalism, like Machiavelli, provides something: the ideal of a perfectly run commonwealth.

Three errors are to be avoided: yet another relativist impasse (Chinese experiences and words are incommensurable); a racialization of China's despotism and the reaffirmation of Kant as the only viable source for universalism; and the consumerist cherry-picking of elements to combine with elements from elsewhere (Celtic, Navajo, Sufi...). All these mask how the intercontinental spaces to be traversed are already structured by racial thresholds and metropolitan privilege.

woman dao of communism

Conceived as oceanic wayfinding rather than homeward footpath, dao is China's greatest contribution to thought. Love and politics are true to the extent they engage an illicit, precarious exercise in wayfinding. Badiou conceives of an eternal "Idea of communism," but this Platonist concept lacks the topological dynamics in the concept of dao. Adopting a term from Deleuze and Guattari, the dao of communism is the transhistorical *diagram* for overcoming injustice, a virtuality coextensive with civilization itself, sensed in the enactment of politics and only describable *post factum*.[85] This diagram is "feminine" insofar as it is radically unfinished and irreconcilable with the false universalism of White Man. As at least one Chinese scholar has argued, Daoism has to be stripped of its gender imagery to become of interest to feminism.[86] Instead of a mega-structuralist and substantive constant, the Daoist conception of femininity is the epistemological remainder wrenched from the patriarchal-bourgeois concealment of how the world works.

If the concept of dao appears conservative or utilitarian its very embedding in eternity can also be used against inherited wisdom. The affirmation that the itinerary of communism *already exists* somewhere out there, even if it is obscure and changes every time we join its eddies, is precisely what is foreclosed by the neoliberal hegemony which has all but usurped the left. The passion of communism comes from its provisional, incremental coherence spurred on by a catastrophic biosphere.

Using the term dao also means that universalization is no departure from the biophysical realm, as tends to be the case for the Kantian, Hegelian, and Lacanian subject. Guattari insisted earlier than most communists that a far-reaching overhaul of the geo-ecological imagination is required.[87] If communist subjectification is called both *ecological* and *feminine*, theosophical connotations of Gaia, womb, oceanicity and so on have to be analyzed not shunned (hence even Lacan's under-analyzed feminine jouissance is open to these connotations). Guattari's "ecosophy" has to be tempered with a nonsubstantivist understanding of becoming: communism's dao is feminine not because it is *yin* but because it is not-all and not-yet, a trajectory away from man's supposed plenitude over and against nature's. The cognitive mapping of the route of anticapitalist universalization is a methodical groping in the dark. While Western ontology can learn much from Dark Studies, femininity is, against most Daoist teaching, a transcendental *political* category first.

Woman's power lies not in maternal care but in nonbeing, partiality, irony, continuous unfolding, the Baroque. As Copjec writes:

> if it is woman who is privileged in Lacan's analysis this is because she remains closer to the truth of being, while man obfuscates this truth through a nostalgic, secondary operation that allows him to maintain a belief in the plenitude of being to come. This is not to say that every woman acts ethically or that no man can ("There are men who are just as good as women. It happens") only that the ethical act is in itself feminine in Lacan's terms. It is woman who is guardian of the not-all of being.[88]

Woman, Freud's dark continent, is indecipherable to man. When assertive, her ambiguities make men queasy. Žižek echoes Irigaray: "behind the multiple layers of masks is nothing; or, at the most, nothing but the shapeless, mucous stuff of the life-substance," and comes close to Wang Bi when going so far as to equate woman with his cherished Hegelian phrase "the night of the world," the terrible inner abyss of human nature which Schelling discovered. Ultimately, "woman is the subject *par excellence*."[89]

Uncharacteristically, Badiou in *Being and Event* introduces an ancient astrological symbol for the "indiscernible," "generic," "unnamable" multiple engaged in a genuinely universalist forcing-open of the status quo: ♀.[90] That he drops the symbol in subsequent work may signal uneasiness in coupling truth as such to femininity. This is a pity. Communism, like all universalism, has always disavowed its sexual cleavage and division of labor. Its separation of public and private, phallic definition of heroism, "cult of personality," occasional predilection for destruction, and stubbornness have prevented a genuine ♀ from forming.

While sexual difference is irrelevant to Badiou's metapolitics his concept of love fully affirms it: "woman and man only ever exist in the field of love."[91] He supplants Lacan's "pessimistic" definition of sex as separation with the positive possibility of committing oneself to working out the fact that in love there is no One (the "fusional" conception). For Irigaray, too, man and woman love each other irreducibly differently and without objectification (I love *to* you). But her hope is that they nonetheless "share a world."[92] Badiou praises love as the ultimate test of becoming-Two, available universally even while soaked with biopower and obscenities.[93] Inaugurated by an existential decision contained in the statement "I love you" following an aleatory encounter, love aims to continue despite the many tribulations revealed in psychoanalysis and the melancholic poetics of its impossibility.

Badiou's anti-fusional position on love can be *politicized.* Feminism shows that sexuation permeates the social field and that politics commences with the deconstruction of the false universality of (bourgeois European) man. I am risking the essentialist notion that in large social formations universality is more accurately graspable by women. Badiou holds that there is a feminine and a masculine approach to the "humanity function" H(x). For the woman-position scientific, artistic, and political truths are only activated by becoming "knotted" through the encounter and proof of the love procedure. For the man-position the four truth procedures occur at the same level, linked through metaphoricity. Hence, unlike for Lacan, for Badiou woman can also universalize precisely because she loves and does not only desire. But he gives no satisfying explanation as to why love is a guarantee of universality for woman alone. What if it is *not love but politics* that knots woman's participation in truth procedures? And is it not men who require a guarantee of being loved before they can access the truths of art, science, religion, and politics?

Guattari notes that

> women are the only authorized repositories of the process of becoming a sexed body. If a man breaks away from the phallic rat race inherent in all power formations, he will become involved in various possible ways in this sort of feminine becoming. Only then can he go on to becoming animal, cosmos, words, colour, music.[94]

With Deleuze he famously hypothesizes:

> It is perhaps the special situation of women in relation to the man-standard that accounts for the fact that becomings, being minoritarian, always pass through a becoming-woman . . . Even blacks, as the Black Panthers said, must become-black. Even women must become-woman.[95]

If becoming-woman is logically prior to racial and class becomings it is not simply because sexuality is the foundation of life. It is also not because woman is yet again projected by man as beacon for his redemption. It is because, freed from polarity, "woman" names the immanent pathway of universality which is hidden from concrete women themselves. Becoming-woman is not fusional or mysterious but strategically participating in the abstract and incomplete idea of emancipation for all, each in their own way.

Of course, revolutionary politics is traditionally sympathetic towards women's emancipation, but this follows from masculinist ways of knowing and plotting. Visiting China in the early 1970s and reading up on Daoism, Julia Kristeva was deeply impressed by the increased mobility of women that the revolution encompassed, but criticizes the prioritizing of industrialization.[96] Mao's anti-Confucianism insisted from the early 1920s that the "four authorities – political, clan, religious and masculine – are the embodiment of the whole feudal-patriarchal ideology and system, and are the four thick ropes binding the Chinese people, particularly the peasants." Nevertheless, sexual difference is strictly superstructural. As he says in 1955: "Genuine equality between the sexes can only be realized in the process of the socialist transformation of society as a whole."[97] For Kristeva as for my feminist universalism, the future of sexual difference lies not in equalizing the sexes but in allowing it to inform every attempt at overall change. Sexual difference is basic (real) like the flow of money. The feminine delivers its productive not-allness *especially* when directed at economic, biological, and geophysical strata. Woman, not man, is the original political animal.

concluding thoughts

In the dialectic of the universal and particular the ontological status of what happens *between* moments of universalization is seldom addressed. If an arc can only retroactively be drawn between/over revolutionary moments to inspire further social struggle, has universalism nothing "to go by" *hic et nunc*? Omitting the topology of actual networks, linkages, and entanglement, post-Hegelian universalism treats isolated event-sites as created-intersected by a universal arriving from an altogether different plane. In Badiou there is a tension between (quite *yang*) ideas of organization and destruction and a (rather *yin*) notion of receptivity to an abrupt moment of clarification. In Žižek's more paradoxical approach the conditions for the revolutionary act only achieve existence *after* it has had its disruptive effects:

> for the famous Leninist theory of the "weakest link" in the chain of Imperialism one does not first ascertain via an objective approach which is this weakest link and then take the decision to strike at this point – the very act of decision *defines* the "weakest link."[98]

Heroic decisionism is based on the self-grounding certitude that our enemy's weaknesses *will be* found without any prior mapping of his networks and our entanglement with them. This is politics without a dao. If action is to be purged of hesitancy, tactical reasoning, and theoretical debate, its epistemic position is almost defensively masculine – cocksure. Why should decisionism necessarily be diluted by the empirical?

As nobody knows beforehand how universalization is to happen we have to trust it can be wrested from the realm of possibility.

Communism as a secularized form of belief is necessary but belief too has to be femininized. Is the post-theological idiom in Badiou and Žižek (and Derrida, Agamben, and many others) the most suitable for arousing political affect? Can the mystical force of conversion-rupture and ensuing fidelity displace the entrenched patrilineal and intolerant biopower of the Judeo-Christian tradition?[99] Even if it is being fruitfully mined for renewing communism, the monotheistic image of thought continues to prevent theorizing how universality could be instantiated in the flesh of billions of humans and a nonhuman, elemental, and purposeless biosphere. The universalities proclaimed in pagan, Asian, and some animist thought are superior by virtue of their already *planetary* yet noncolonial scope.

Ecology hardly enters the conceptual frames of post-Hegelians. Although nature is of concern to Žižek by way of Schelling, it is as a primeval, deterministically apocalyptic realm, not complexifying or surprising on its own accord. The German framework proves excellent for thinking ethics and politics but it cannot think climatology and geology without thinking a subject already thinking them. In the Deleuzian as well as Daoist conception, nature stands very much by itself and severely constrains how humans inhabit "it." Thought does not liberate itself from, but is a secretion of an irrecuperably obscure biophysical backdrop. Hinging this obscurity on the unconscious or signifier alone is still a kind of humanism, preventing passages across epistemic thresholds that the human predicament under climate change calls for.

Irigaray provides inspiration for a feminine biocommunism. She correctly advocates greater representation of women in politics, religion, art, science, and philosophy, but foresees the transformation of their institutional matrices as longer-term projects. Lured into state desire, Irigaray reduces politics to citizenship and rights, while feminist universalism commences *between* states and cultures. How else to confront the trafficking of Burmese girls and boys for white men in Bangkok, the use of rape as terror strategy amongst refugee populations in Sudan or Bosnia, "femicide" in Mexican border cities, white fantasies about Arab women and Arab fantasies about white women, or the sleek machismo of business travel?

Instead of channeling spiritual energies towards peaceful coexistence, universalism should now carry on the transhistorical, transcontinental diagram of revolution grasped and reinvented by the women and men of Cairo in January 2011, even if it was not sustained and focused as "communist." That woman is the Way of communism does not mean she is the communist *subject* – which will always be the poor masses and their intellectual allies of all sexes – but that this subject is to be guided transcendentally by the not-all, the to-come, and the multiplicitousness that is formally hers. The proletariat has to become-woman for the exact reason that capital and environmental racism are masculinist. To paraphrase Spinoza, what is least known in our world is what (a) woman can do.

notes

The epigraph is from *The Book of Chuang Tzu*, trans. M. Palmer with E. Breuilly (London: Penguin, 2006) 236–37. Written in the fourth century BCE.

1 Luce Irigaray, *Sexes and Genealogies*, trans. Gillian C. Gill (New York: Columbia UP, 1993) 140.

2 Idem, *I Love to You: Sketch for a Felicity in History*, trans. Alison Martin (London: Routledge, 1996); idem, *Democracy Begins between Two*, trans. Kirsteen Anderson (London: Athlone, 2001).

3 Idem, *The Way of Love*, trans. Heidi Bostic and Stephen Pluháček (London: Continuum, 2002).

4 Gilles Deleuze and Félix Guattari, *A Thousand Plateaus: Capitalism and Schizophrenia*, trans. Brian Massumi (Minneapolis: U of Minnesota P, 1987) 176ff.

5 Naomi Schor, "French Feminism is a Universalism," *Differences* 7.1 (1995): 15–38.

6 Alain Badiou with Nicolas Truong, *Éloge de l'amour* (Paris: Flammarion, 2009) 52–56.

7 Félix Guattari, "The Micro-Politics of Fascism" in *Molecular Revolution: Psychiatry and Politics*, trans. Rosemary Seed (London: Penguin, 1984) 217–32. Though Guattari himself does not espouse

universalism, his transversal communism nevertheless aims at trespassing all prior divisions of humanity.

8 Félix Guattari and Antonio Negri, *Communists Like Us: New Spaces of Liberty, New Lines of Alliance*, trans. Michael Ryan (New York: Semiotext(e), 1990).

9 Alain Badiou, *The Communist Hypothesis* (London: Verso, 2010).

10 Irigaray, *I Love to You* 47.

11 The concern that Irigaray neglects trans-, inter-, and queer sexualities downplays decades of debate on the intractable *failures* of actual sexual difference to stand up to its posited binarity. Underestimating the work that most humans put in conforming to sexual dimorphism, many queer theorists and interpretations of Deleuze and Guattari, even when they are correct in denouncing heteronormativity, postulate too cheerfully a realm of betweenness as new global destination.

12 Penelope Deutscher, *A Politics of Impossible Difference: The Later Work of Luce Irigaray* (Ithaca, NY: Cornell UP, 2007).

13 Margaret Whitford, "Irigaray, Utopia, and the Death Drive" in *Engaging with Irigaray: Feminist Philosophy and Modern European Thought*, eds. Carolyne Burke, Naomi Schor, and Margaret Whitford (New York: Columbia UP, 1994) 381; Luce Irigaray, *Speculum of the Other Woman*, trans. Gillian C. Gill (Ithaca, NY: Cornell UP, 1985), which appeared in French in 1974.

14 Gilles Deleuze and Félix Guattari, *What is Philosophy?*, trans. Graham Burchell and Hugh Tomlinson (New York: Columbia UP, 1994) 10.

15 Alberto Toscano, "The Politics of Abstraction: Communism and Philosophy" in *The Idea of Communism*, eds. Costas Douzinas and Slavoj Žižek (London: Verso, 2010) 197.

16 See, for example, Slavoj Žižek, "Plagiarizing from the Future," *Lacanian Ink* 36 (2010): 148–61.

17 Slavoj Žižek and Mao Tse-tung, *Slavoj Žižek Presents Mao: On Practice and Contradiction* (London: Verso, 2007) 42.

18 Elizabeth Grosz, "Sexual Difference and the Problem of Essentialism" in *The Essential Difference*, eds. Naomi Schor and Elizabeth Weed (Bloomington: Indiana UP, 1994) 82–97.

19 See Elizabeth Grosz, *Sexual Subversions: Three French Feminists* (Sydney: Allen, 1989).

20 Jacques Lacan, "The Subversion of the Subject and the Dialectic of Desire" in *Écrits*, trans. Bruce Fink (New York: Norton, 2002) 671–702.

21 Ibid. 695.

22 A central source on the political intricacies of the dialectic is the discussion between Laclau, Žižek, and Judith Butler in *Contingency, Hegemony, Universality: Contemporary Dialogues on the Left* (London: Verso, 1999).

23 Ernesto Laclau, *Emancipation(s)* (London: Verso, 1996) 52–54.

24 Ibid. 57.

25 Linda M.G. Zirelli, "This Universalism Which is Not One," *Diacritics* 28.2 (1998): 3–20.

26 Laclau, *Emancipation(s)* 33.

27 Joan W. Scott, "Universalism and the History of Feminism," *Differences* 7.1 (1995) 7.

28 Alain Badiou, *Ethics: An Essay on the Understanding of Evil*, trans. Peter Hallward (London: Verso, 2001).

29 See idem, *The Rational Kernel of the Hegelian Dialectic*, trans. Tzuchien Tho (Melbourne: re:press, 2011), an early essay reconstructing Hegel via Maoism.

30 Peter Hallward, "Translator's introduction," Badiou, *Ethics* xxviii.

31 Irigaray is easily distinguished from socialist feminism. First, her politics derives from the universality of sex not capital. Second, Marx is only one in a series of male conversation partners, including Plato, Merleau-Ponty, and Heidegger, of little relevance to socialist feminism. Third, the spaces where creativity emerges for Irigaray are conventionally called secondary in Marxism: language, touch, the imagination, etc.

32 Luce Irigaray, *I Love to You*; idem, *Thinking the Difference: For a Peaceful Revolution*, trans. Karen Montin (London: Athlone, 1994).

33 Sebastian Budgen, Stathis Kouvelakis, and Slavoj Žižek, eds., *Lenin Reloaded: Toward a Politics of Truth* (Durham, NC: Duke UP, 2007); Yannis Stavrakakis, *Lacan and the Political* (London: Routledge, 1999).

34 Alain Badiou, *Saint Paul: The Foundation of Universalism*, trans. Ray Brassier (Stanford, CA: Stanford UP, 2003).

35 Fredric Jameson, "Cognitive Mapping" in *The Jameson Reader*, eds. Michael Hardt and Kathi Weeks (Malden, MA: Blackwell, 2000) 277–87.

36 My stupidity, too, will at some moment become clear.

37 See James Penney, "(Queer) Theory and the Universal Alternative," *Diacritics* 32.2 (2002): 3–19; Nina Power, *One Dimensional Woman* (London: Zero, 2009); Jasbir Puar, *Terrorist Assemblages: Homonationalism in Queer Times* (Durham, NC: Duke UP, 2007).

38 This essay focuses on universalism in politics even though science, art, pedagogy, the media, and travel also are in dire need of universal-feminism.

39 In Dorothea Olkowksi's *The Universal (in the Realm of the Sensible)* (New York: Columbia UP, 2007) the subversive agendas of feminism and universalism unfortunately get lost in the sheer expanse of Deleuzian and Kantian musings on phenomenology, cosmology, poetry, and logic.

40 Luce Irigaray, *Between East and West: Between Singularity and Community*, trans. Stephen Pluháček (New York: Columbia UP, 2002) 48.

41 For universal-feminism more comparison would be in order. Sexual dualisms are ubiquitous and became especially abstract in Mesoamerican cosmologies. The Mexican dialectic, though also emanating from feudal-military empire, places more emphasis on the death drive, possibly because it remains rooted in hallucinogenic rites. See Sylvia Marcos's Irigarayan interpretation of Aztec sexualities in *Taken From the Lips: Gender and Eros in Mesoamerican Religions* (Leiden: Brill, 2006). The question is indeed why European philosophy did not resort to male and female vital principles more often.

42 Irigaray, *Sexes and Genealogies* 57.

43 The orientalization of the Orient is now associated with the Wade-Giles spelling and I use the pinyin system throughout – Daoism instead of Taoism, Mao Zedong instead of Tse-tung. Daoism includes its popularization in the West, the Chinese diaspora, and post-Mao Chinese capitalism. Though intercultural efforts like Fritjof Capra's *The Tao of Physics: An Exploration of the Parallels between Modern Physics and Eastern Mysticism*, 25th anniversary ed. (Boston: Shambhala, 2000) deserve close analysis, it is also evident that New Age interpretations of the Way form an ideological accompaniment to laissez-faire economics and the self-cultivation industry. The aura emanating from ☯ indexes the fact that a metaphysic of sexual difference is widespread in consumer capitalism; unsurprisingly ☯ is most often encountered in sex manuals.

44 The Dark as expounded here approximates a materialist adoption of the null-set (∅) which uniquely accompanies every other set, what Badiou calls the "subtractive suture of being" in *Being and Event*, trans. Oliver Feltham (London: Continuum, 2005) 66ff. Nonetheless, void is fully real insofar as it is to-become-physical, interstitial to being. Particle physics is helpful here; see Frank Close, *The Void* (Oxford: Oxford UP, 2007).

45 Confucius, *The Analects*, trans. Raymond Dawson (Oxford: Oxford UP, 1993) 12.12.

46 Kenneth Dean and Brian Massumi argue in *First and Last Emperors: The Absolute State and the Body of the Despot* (Brooklyn: Autonomedia, 1992) that in its intolerance of "floating elements" – scholars, ascetics, salespeople, beggars, outlaws, lower clerks, sex workers – the First Emperor invented the diagram of the perfect state, still discernible in the corporatist United States. One major difference between the American and Chinese empires is, of course, that the latter enfolded professional philosophy into its very functioning.

47 Sarah Allan, *The Way of Water and Sprouts of Virtue* (Albany: State U of New York P, 1997).

48 Laozi, *Daodejing*, trans. Edmund Ryden (Oxford: Oxford UP, 2008) chapter 1. Cf. Stephen Mitchell, *Dao Te Ching: A New English Version* (New York: Harper, 1998); Richard G. Wagner, *A Chinese Reading of the Daodejing: Wang Bi's Commentary on the* Laozi *with Critical Text and Translation* (Albany: State U of New York P, 2003). Subsequent chapter references are given with translator.

49 Mitchell's freer wording has a similar perverse ring to it: "The Dao is called the Great Mother: empty yet inexhaustible, it gives birth to infinite worlds. It is always present within you. You can use it any way you want."

50 Brian Massumi, "Translator's Foreword: Pleasures of Philosophy" in Deleuze and Guattari, *A Thousand Plateaus* xi–xii.

51 Dean and Massumi, *First and Last Emperors* 83. Dean and Massumi let Daoism off the hook.

52 A.C. Graham, *Yin-Yang and the Nature of Correlative Thinking* (Singapore: Institute of East Asian Philosophies, 1986).

53 James Miller, *Daoism* (Oxford: OneWorld, 2003) 51.

54 Irigaray, *Speculum* 237.

55 Idem, *Marine Lover of Friedrich Nietzsche*, trans. Gillian C. Gill (New York: Columbia UP, 1991); idem, "The 'Mechanics' of Fluids" in *This Sex Which is Not One*, trans. Catherine Porter (Ithaca, NY: Cornell UP, 1985) 106–18.

56 Irigaray, *Marine Lover* 37.

57 Alison Stone, *Luce Irigaray and the Philosophy of Difference* (Cambridge: Cambridge UP, 2006).

58 Elizabeth Grosz, *Becoming Undone: Darwinian Reflections on Life, Politics, and Art* (Durham, NC: Duke UP, 2011).

59 I take the liberty of combining the translations of Mitchell, Ryden, and Wagner.

60 Wagner, *A Chinese Reading* 266.

61 Richard G. Wagner, *Language, Ontology, and Political Philosophy in China: Wang Bi's Scholarly Explorations of the Dark* (Xuanxue) (Albany: State U of New York P, 2003) 69–70.

62 Ibid. 143–44.

63 Dark vitalism is what philosophy blogger Naught Thought calls his brand of speculative realism.

64 Slavoj Žižek, *For They Know Not What They Do: Enjoyment as a Political Factor*, 2nd ed. (London: Verso, 2008) xxvi.

65 Miller, *Daoism* 164.

66 Slavoj Žižek, *The Parallax View* (Cambridge, MA: MIT P, 2006) 7.

67 Jacques Lacan, *D'un discours qui ne serait pas du semblant. Le Séminaire de Jacques Lacan Livre XVIII, 1971*, ed. Jacques-Alain Miller (Paris: Seuil, 2007). Guy Flecher has compiled all Lacan's references to Chinese philosophy and makes them converge on <www.lacanchine.com>. On sexuation see Jacques Lacan, *Encore: The Seminar of Jacques Lacan Book XX. On Feminine Sexuality, the Limits of Love and Knowledge, 1972–1973*, ed. Jacques-Alain Miller; trans. Bruce Fink (New York: Norton, 1998).

68 Lacan, *D'un discours qui ne serait pas du semblant* 65–71.

69 Bruce Fink's translation of *côté* (side, in the table of sexuation) as "pole" is unfortunate. See his note in Lacan, *Encore* 8 n. 32.

70 Ibid. 73.

71 Joan Copjec, "*m/f*, or Not Reconciled" in *The Woman in Question: m/f*, eds. Parveen Adams and Elizabeth Cowie (Cambridge, MA: MIT P, 1990) 17.

72

> [The supplement] adds only to replace. It intervenes or insinuates itself *in-the-place-of*; if it fills, it is as if one fills a void . . . : whether it adds or substitutes itself, the supplement is *exterior*, outside of the positivity to which it is super-added, alien to that which, in order to be replaced by it, must be other than it. Unlike the *complement*, dictionaries tell us, the supplement is an "*exterior* addition." (Jacques Derrida, *Of Grammatology*, trans. Gayatri C. Spivak (Baltimore: Johns Hopkins UP, 1997) 145)

73 Luce Irigaray, *Je, Tu, Nous: Toward a Culture of Difference*, trans. Alison Martin (London: Routledge, 1993).

74 Idem, *To Be Two*, trans. Monique Rhodes and Marco F. Cocito-Monoc (New York: Routledge, 2001).

75 Emphasizing jealousy, misogyny, emasculation, etc., the usual explanations of Islamic fundamentalism conceal how Western interests produce the socioeconomic conditions for political radicalization. Inversely, Islamophobia derives primarily from anxieties about property and borders, not from imagining a diabolical other possessing a mysterious fullness of being. More than the reproduction of stereotypes, the ideological problem here is the obscuring of material relations. Psychoanalysis falls into an ideological trap when it treats racism and nationalism as individual pathologies – Žižek's figure of the

anti-Semite – instead of intrinsic to the state, division of labor, medicine, and policing.

76 Michel Foucault, "Right of Death and Power over Life" in *The History of Sexuality, Volume I: An Introduction*, trans. Robert Hurley (New York: Random, 1978) 133–60.

77 My argument combines Janet Abu-Lughod, *Before European Hegemony: The World System A.D. 1250–1350* (New York: Oxford UP, 1989); Samir Amin, *Eurocentrism*, 2nd ed. (New York: Monthly, 2010); David Harvey, *The New Imperialism* (Oxford: Oxford UP, 2003); and Denise Fereira da Silva, *Toward a Global Idea of Race* (Minneapolis: U of Minnesota P, 2007). One could understand primitive accumulation as "primitive" not in Marx's chrono-logical sense but in that it reveals the savagery of European modernity.

78 Kalpana Seshadri-Crooks, *Desiring Whiteness: A Lacanian Analysis of Race* (London: Routledge, 2000). Cf. Ann Laura Stoler, *Race and the Education of Desire: Foucault's History of Sexuality and the Colonial Order of Things* (Durham, NC: Duke UP, 1995).

79 Helena P. Blavatsky, *The Secret Doctrine* (Madras: Theosophical, 1978).

80 More on this paradox in my *Psychedelic White: Goa Trance and the Viscosity of Race* (Minneapolis: U of Minnesota P, 2007).

81 Francis Gray, *Jung, Irigaray, Individuation: Philosophy, Analytical Psychology, and the Question of the Feminine* (New York: Routledge, 2008). Incidentally, the commonalities of Jung and Daoism have also often been pointed out. See the biographical sketch in David H. Rosen, *The Tao of Jung: The Way of Integrity* (New York: Penguin, 1996).

82 Joseph Goebbels quoted in Andrew Samuels, "New Material Concerning Jung, Anti-Semitism, and the Nazis," *Journal of Analytical Psychology* 38.4 (1993) 465.

83 Cf. James E. Goggin and Eileen Brockman-Goggin, *Death of a "Jewish Science": Psychoanalysis in the Third Reich* (West Lafayette, IN: Purdue UP, 2001), especially chapter 5.

84 Slavoj Žižek, "Mao Tsetung: The Marxist Lord of Misrule" in *On Practice and Contradiction* 9–10. The ancient philosophy that Mao did loosely relate to was not Daoism but Legalism. Badiou subsequently comes to celebrate Legalism as precursor of Robespierre and Mao, downplaying its totalitarianism in favor of its antimoralist formalism. *Logics of Worlds: Being and Event 2*, trans. Alberto Toscano (London: Continuum, 2009) 20ff. Han Fei (?280–233 BCE) compiled Legalist thought *only* to create an empire based on a paranoid penal system. Han Feizi, *Basic Writings*, ed. and trans. Burton Watson (New York: Columbia UP, 1964). Bringing Daoism into the (Hegelian) equation, Žižek also does not consider this absolute predominance of state desire:

> The ideological constellation in ancient China was dominated by the opposition between Confucianism (reliance on traditional customs, authority, and education) and Taoism (spontaneous self-enlightenment) – with the uncanny *third* position of "legalists" rehabilitated by Mao Zedong, partisans of egalitarian revolutionary terror. In our perception, today's ideological constellation is determined by the opposition between neoconservative fundamentalist populism and liberal multiculturalism – both parasitizing on each other, both precluding any alternative to the system as such. (*Parallax View* 349)

There are Confucian echoes in Bushite neoconservatism but far more so of Legalism, whose terror is not uncanny or egalitarian but ruthlessly simple and authoritarian. Meanwhile Daoism could arguably resist the pull of liberal multiculturalism insofar as it sticks to the paradox, negativity, and anti-imperialism propounded in the *Daodejing*.

85 Deleuze and Guattari, *A Thousand Plateaus* 140–43.

86 Karyn Lai, "The *Daodejing*: Resources for Contemporary Feminist Thinking," *Journal of Chinese Philosophy* 27.2 (2002): 131–53.

87 Félix Guattari, *The Three Ecologies*, trans. Ian Pindar and Paul Sutton (London: Athlone, 2000).

88 Joan Copjec, *Imagine There's No Woman: Ethics and Sublimation* (Cambridge, MA: MIT P, 2003) 7.

89 Slavoj Žižek, *The Metastases of Enjoyment: Six Essays on Woman and Causality* (London: Verso, 1994) 150, 143.

90 Badiou, *Being and Event* 256ff.

91 Idem, "What is Love?" in *Conditions*, trans. Steve Corcoran (London: Continuum, 2008) 196.

92 Luce Irigaray, *Sharing the World* (London: Continuum, 2008).

93 See Badiou with Truong, *Éloge de l'amour.*

94 Guattari, *Molecular Revolution* 234.

95 Deleuze and Guattari, *A Thousand Plateaus* 291.

96 Julia Kristeva, *About Chinese Women*, trans. Anita Barrows (New York: Boyars, 1977).

97 Mao Tsetung, *Quotations from Chairman Mao Tsetung* (Peking: Foreign, 1972) 294, 297.

98 Žižek, *For They Know Not What They Do* 192.

99 Žižek's voracious universalism can be careless, as when he makes a slip while arguing for the superiority of Christianity:

> while in a tragedy, the individual actor represents the universal character he plays, in a comedy, he immediately *is* this character. The gap of representation is thus closed, exactly as in the case of Christ who, in contrast to previous pagan divinities, does not "represent" some universal power or principle (as in Hinduism, in which Krishna, Vishna [*sic*], Shiva, and so on, all "stand for" certain spiritual principles or powers – love, hatred, reason): as this miserable human, Christ directly *is* God. (*Parallax View* 105)

As far I know Hindu divinities do not represent but *are* complex literal embodiments, not of anthropic conditions like love or reason but of cosmological principles and mythic narrative. Krishna is the most important avatar of Vishnu, who ensures the maintenance of things; Brahma is their Creator, Shiva their Destroyer. Except for some androgynous depictions of Shiva all three are clearly male. In Vedanta philosophy Brahma, Vishnu, and Shiva form the quasi-monotheistic unity of the One-Absolute, the unsexed Brahman.

Does Žižek's slip hint at an unconscious femininization of polytheism for being naive and local? He later juxtaposes Che Guevara to what he calls the "Oriental" solution to the antinomy of renunciation and struggle exemplified in the brainwashed Zen warrior. "Freedom is not a blissfully neutral state of harmony and balance, but the very violent act which disturbs this balance. Buddhist (or Hindu, for that matter) all-encompassing Compassion must be opposed to Christian *intolerant, violent* Love" (*Parallax View* 282). Of course belief systems are more polyvalent than such an opposition implies. More important for universalism is to avoid the masculinism of an evangelical position basing belief in Jesus and Che as sacrificial *exceptions* to ordinary manhood to be approximated across the world.

rosi braidotti

AFTERWORD
complexity, materialism, difference

Readers of this important and agenda-setting issue of *Angelaki* are confronted by a number of key questions concerning the scientific and social status of the concepts of complexity, materialism, difference, and the process of subject formation. How has the historical context of advanced capitalism altered our understanding of embodiment, complexity, and interrelationality? How has materialism changed in response to the different understandings of "matter" induced by the current scientific advances? How do they affect specifically bodily or corporeal materialism? How do the new notions of "matter" affect critiques of biopower on the one hand and new forms of vital politics on the other? How does the practice of the political shift in response to these? And how is the political and theoretical project of sexual difference inscribed in this new context?

the perverse materiality of advanced capitalism

Advanced capitalism is a difference-engine that promotes the marketing of pluralistic differences and the commodification of the existence, culture, and genetic codes of multiple "others" for the purpose of consumerism. The global system of postindustrial and newly industrializing worlds produces scattered and poly-centered yet always profit-oriented power relations which function not so much by binary oppositions but in a fragmented and all-pervasive manner. The rhizomic or web-like structure of contemporary power, however, does not alter fundamentally its terms of application. If anything, power relations in globalization are more ruthless than ever. Advanced capitalism also engenders, therefore, a problematic form of mobility because the much-celebrated practice of "free circulation" pertains exclusively to the domain of goods and commodities, provided they guarantee maximum profit. People do not circulate nearly as freely.[1] This perverse variation on nomadism expresses a logic of economic exploitation that equates capitalist flows and flux with the circulation of commodities.[2] It also produces a notion of pan-humanity as united by shared threats and dangers in what has become a global risk society.[3] If global communication networks can engender a positive form of proximity, epidemics, environmental catastrophes, and militarization construct a negative form of planetary interconnection.

Considering the extent to which contemporary economies depend on the commodification of life itself, an instrumental form of posthumanism emerges from the very post-anthropocentric opportunism of advanced capitalism. The biogenetic dimension of advanced capitalism is such that it is not only genocentric[4] but also ruthlessly and structurally unjust. Deleuze and Guattari analyzed this situation in terms of capitalism as a conflict between, on the one hand, the rising demands for subjective singularities and, on the other, the conservative reterritorialization of desires for the purpose of commercial profit.[5] This internally contradictory movement achieves the doubly disastrous effect of reasserting consumerist liberal individualism as the unquestionable standard for subject formation, while reattaching it to the multiplying and spinning power of contemporary technologically mediated global economies.

Technologies – more specifically information and biotechnologies – are intrinsic to the social and discursive structures of postindustrial societies. They are also central to the perverse shift in the political economy of difference: the proliferation of quantitative differences – multiples of One – on a planetary scale. Access to and participation in the new high-tech world are, however, unevenly distributed worldwide, with gender, age, and ethnicity acting as the major axes of negative differentiation.[6]

Two major notions get dislocated in this political economy of spinning differences: the first is our understanding of matter itself and the second the relation between the organic and the technological artifact.

Firstly, the question of technology raises its multifaceted head once again and forces us to question what exactly is going on between technology and the space that is already predesignated as the interval between bodies. Heidegger both humanizes technology and makes it part of the human capacity for ecstatic ex-centricity (standing beside oneself in conscious self-reflexivity) which he sees as constitutive of the human subject.[7] Postmodernity is the historical time when such ontological distinctions collapse. Our historical context plunges us, therefore, in an extended post-Heideggerian limbo, caught between spectral echoes of posthuman humanisms and nostalgic longings for new universals.

We are caught in a perverse materiality, in that our era is marked by a new and perversely fruitful alliance with technology which stresses the proximity, familiarity, and increased intimacy between the human and the non-human technological universe. In the cyberuniverse that we inhabit a rather complex bond of mutual dependence has emerged between the flesh and the machine. This engenders some significant paradoxes, especially when it comes to the human body: the corporeal site of subjectivity is simultaneously denied, in a fantasy of escape, and strengthened or re-enforced. Bodily materialism is altered accordingly.

Deleuze's anti-essentialist vitalism and emphasis on complexity offer a rigorous call for rethinking human embodiment in a manner that is coextensive with our technological habitat. Deleuze and Guattari show that established ideas of the organic and the mechanical are equally molar, or sedentary. In opposition to the holistic, neo-Romantic, and anxious view of the mechanical world, they defend a molecular, machinic one, which is about becomings, without ultimate purpose or finality. A kind of generalized "becoming-bachelor-machine."[8]

This is an important cautionary note because, as Keith Ansell Pearson argues, some grand narratives have come back into fashion that tend to be deterministic and evolutionary in a naïve and oddly old-fashioned way: "A new mythology of the machine is emerging and finds expression in current claims that technology is simply the pursuit of life by means other than life."[9] This simplistic and reductive reading of the transformations currently at work in our global system reveals great conceptual poverty. A hierarchical fantasy of vertical perfectibility, a technologically mediated quest for immortality and for disciplined and acquiescent subjects, has gained widespread currency, which betrays the nomadic potential of contemporary science.[10] In opposition to this master narrative, which corresponds to what Donna Haraway calls "the informatics of domination,"[11] feminist and nomadic matter-realist philosophers stress the relevance of materialist, vital, and complex philosophies of becoming, as an alternative conceptual framework, in the service of a sustainable future.

matter-realism and vital politics

Secondly, about the new understandings of "matter": contemporary capitalism is "bio-political" in that it attempts to control all that lives: it has already turned into a form of generalized biopiracy aiming to exploit the generative powers of women, animals, plants, genes, and cells. This means that human and anthropomorphic others are relocated in a continuum together with non-anthropomorphic or "earth" others. The categorical distinction that separated the Human from his (the gender is no coincidence) naturalized others has shifted, taking the humanist assumptions about what constitutes the basic unit of reference for the "human" into a spin.

The contemporary social imaginary in relation to these changes swings between euphoric techno-transcendentalism and paranoid technophobia.[12] This manic-depressive associative nexus results in negative representations of different "others" – the techno-monsters of late, postindustrial popular culture. An essential element of their monstrosity is an element of hybridity – that is to say the blurring of categorical distinctions or constitutive boundaries. Primary blurrings occur between different sexes, races, and species – the human/the animal, the organic/the inorganic/technological.

The blurring of these categorical divides between self and others creates a sort of heteroglossia, a colossal hybridization that ends up locating "Life itself" as non-human agent at the center of scientific and political debate.[13] After so much emphasis on the linguistic and the cultural, an ontology of presence replaces textual or other deconstruction.

I refer to the neorealist practices of bodily materialism as "matter-realism,"[14] radical neo-materialism, or posthuman feminism. One of the main reasons to advocate these is the changing conceptual structure of materialism itself through the impact of contemporary biogenetics and information technologies. The Spinozist switch to a monistic political ontology stresses processes, vital politics, and non-deterministic theories of evolution – as argued brilliantly by Elizabeth Grosz in this and other volumes. Politically, the emphasis falls accordingly on the micropolitics of relations, as a posthumanist ethics that traces transversal connections among material and symbolic, concrete and discursive, lines and forces. Transversality actualizes an ethics based on the primacy of the relation, of interdependence, which values non-human or a-personal Life. This is what I call *zoe*-posthuman politics.[15]

The most striking feature of the current scientific redefinition of "matter" is the wider acceptance granted to the shift of emphasis in thinking about difference from binaries to rhizomatics. Critical theory needs to look carefully at the dislocation of the dialectical relationships between the traditional axes of difference: sexualization/racialization/naturalization and attempts to come to terms with this challenge. It can also be described as a sort of "anthropological exodus" from the dominant configurations of the human[16] – a massive hybridization of the species which topples the Human from the sovereign position it has enjoyed for so long. This standard is posited in a universal mode as Man, but this pseudo-universal has been widely criticized precisely because of its partiality.[17] Universal Man, in fact, is implicitly assumed to be masculine, white, urbanized, speaking a standard language, heterosexually inscribed in a reproductive unit and a full citizen of a recognized polity.[18]

Brian Massumi refers to this phenomenon as "Ex-Man," "a genetic matrix embedded in the materiality of the human" and as such undergoing significant mutations: "species integrity is lost in a bio-chemical mode expressing the mutability of human matter."[19] Haraway sums up the current predicament as follows: "This is Man the taxonomic type become Man the brand."[20] The perverse economy of our times challenges post-anthropocentrism. What emerges is the vital politics of posthuman life.

What matters for materialism today is consequently the concept of matter itself.[21] The switch to a monistic political ontology stresses processes, vital politics, and non-deterministic evolutionary theories.[22] French epistemology has a long and glorious history of "enchanted materialism," that is to say a nonreductive understanding of the material roots of human consciousness and of scientific knowledge production. The respective work of Foucault's teacher Georges Canguilhem and of Deleuze's teacher Gaston Bachelard alone bears testimony to the enormous influence of this epistemological tradition. The former argues that a sort of primitive anthropomorphism is implicit in the technological artifact and is written into the whole history of technology. The latter argues that embodiment functions, in this tradition of thought, as a crucial category which also enlists the forces of the imagination in the service of knowledge production.

This assertion of the primacy of the living organism over the technological implies that the technological other imitates the organic self. All technologies consequently become biotechnologies. The great advantage of this technophilic biophilosophy is that it paves the road for rethinking the symbiotic relationship between the human and the technological in productive and non-essentialist ways. This project will be implemented with brilliant originality by one of Canguilhem's American students – Donna Haraway – who introduced an affirmative and subversive form of technophilia into contemporary social and cultural theory and the notoriously technophobic feminist community. Technology is approached not only as the expression of the desire for mastery but also an object of desire, curiosity, and affective involvement.

The technobodies of late postindustrial societies can and should be understood in the light of the increasingly complex aggregation of information that contemporary science has been able to provide about them. With reference to molecular biology, genetics, and neurology – to mention just a few scientific fields – the body today can and should be described adequately as a sensor with serious credibility and an integrated site of information networks. It is a messenger carrying thousands of communication systems: cardiovascular, respiratory, visual, acoustic, tactile, olfactory, hormonal, psychic, emotional, erotic, etc., etc. Coordinated by an inimitable circuit of information transmission, the body is a living recording system, capable of storing and then retrieving the necessary information to process it at such speed that it can react "instinctively." Fundamentally prone to pleasure, the embodied subject tends towards the recollection and repetition of experiences which pleasure has "fixed" psychically and sensually within the subject (to re-member, after all, is to re-peat and repetition tends to favor that which gave joy and not that which gave pain). The body is not only multifunctional but also in some ways multilingual: it speaks through temperature, motion, speed, emotions, excitement that affect the cardiac and other rhythms.

To account for the embodied self in keeping with contemporary understandings of matter and of complexity exemplifies the type of vitalistic materialism which the contributors to this special issue have illustrated brilliantly. The strategic position of sexual difference is inscribed at the heart of the perverse conflicts of interests of postindustrial technosciences. This special issue makes a significant contribution to the debate by striking a rigorous and coherent note of resistance against the neutralization of difference that is induced by the perverse materiality and the tendentious mobility of advanced capitalism. This system engenders a deliberate blurring of dichotomous differences, which, however, does not resolve or improve the mechanisms of power and in many ways exacerbates them. Taking a firm stand against these trends, this issue of *Angelaki* reiterates the notion that although – significantly enough – it is still up to the minorities to try to ground, i.e., to provide an accountable location for, alternative views of the subjects and hence also new images of thought, the issue of sexual difference in the formation of subjectivity is more timely than ever. Trust women, gays, and other alternative forces, with their historically leaky bodies,[23] to both reassert the powers and enhance the potentiality of the "wetware."[24] Let us trust a little more to contemporary evolutionary theories and biogenetics, however, to extend this prerogative to multiple non-human agents as well.

sexual difference as the principle of not-one

The matter-realist turn has important implications for the discussion of sexuality and gender. Contrary to the social-constructivist methodology of gender and queer theory, contemporary sexual difference theory proposes a vital politics of embodiment as becoming: sexuality beyond gender. Whereas queer theory in line with the linguistic turn locates the heterosexual discursive matrix at the core of its analyses of power,[25] in the vitalist tradition we return to the main insights about the polymorphous and perverse structure of human sexuality, to which we must add, however, the generative powers of female embodiment. In this perspective, gender is just one historically contingent mechanism of capturing the multiple potentialities of the body, including their generative or reproductive capacities. To turn it into *the* transhistorical matrix of power is quite simply a conceptual error and a political mystification. In a monist political economy, power is a not a static given but a complex strategic flow of effects which call for a pragmatic politics of intervention and the quest for sustainable alternatives.[26]

Thus, Irigaray stresses the potency of female bodies as sites of the material, maternal power of engendering the future. She also emphasizes the symbolic and material costs of the "theft" of women's sexuality through the sacrificial ontology of a phallocentric system that requires the exchange of women to fuel its socio-symbolic structures. The emphasis thus falls on the specificity of women's own sexual economy.[27] Deleuze and Guattari, on the other hand, broaden the scope of the discussion by stressing that the gender system captures the complexity of human sexuality in a binary machine that privileges heterosexual family formations and literally steals all other possible corporealities from us. As a result we no longer know what sexed bodies can do.

Both Irigaray and Deleuze challenge queer theory's reductive rendition of the original foreclosure of the first love object – the mother – and of the sexual complexity that marks the polymorphous and perverse structure of human sexuality. Both engage, in different but powerful ways, with the unconscious or transhistorical and transpersonal carnal elements that are involved in the process of capture or theft from the originary polymorphous sexual body. What is emerging more clearly in current discussions about sexuality is that, whereas most queer theory is solidly ensconced in social-constructivist methods and political strategies, matter-realist thinkers affirm and explore the ontological aspects of sexuality and sexual difference, and not only its representational elements.

As a consequence, matter-realist or vitalist feminism, resting on a dynamic monistic political ontology, shifts the focus away from the sex/gender distinction, bringing sexuality as process into full focus. A first problem concerns the irrelevance of the category "same-sex" to account for the complex and multiple affects generated in the relation between two human beings. The redundancy of the sex/gender distinction for a feminist philosophy of the subject was noted in the late 1980s and early 1990s by English-speaking feminists working on Continental philosophy such as Judith Butler, Moira Gatens, Elizabeth Grosz, and me.

Contemporary feminist philosophers argue the same case in line with Deleuzian ontology. For instance, Claire Colebrook stresses that sexual difference is not a problem that needs a solution but a productive location to start from.[28] Patricia MacCormack draws attention to the need to return to sexuality as a polymorphous and complex force and to disengage it from both identity and all dualistic oppositions. She looks for subversion not in counter-identity formations but rather in pure dislocations of identities via perversion of standardized patterns of interaction.[29] The ethics of becoming is an ethology of the forces that propel the subject to overcome both forms and categories, deterritorializing all identities on its line of flight. This means by extension that sexuality is a force, or constitutive element, that is capable of deterritorializing gender – and the institutions that reproduce it.

Sexual difference, from being a boundary-marker, has become, therefore, a threshold for the elaboration and the expression of multiple differences, which extend beyond gender but also beyond the human. The emergence of the earth, and of "earth-others" as political subjects, is the surest indicator of this shift of perspective. I want to read this with the ecophilosophy of nomadism, as an attempt to rethink in a materialist manner the intricate web of interrelations that mark the contemporary relationship to one's environment.

Notions such as sexual difference as the principle that activates what I have defined as the "virtual feminine"[30] and Deleuze's theory of becoming-woman/-animal/-machine/-imperceptible provide new ontological grounds to recast a form of subjectivity which no longer assumes the unitary, self-evident subject of modernity or the essentialized subject of "standpoint" feminism. This is, rather, a non-unitary, multiple, complex subject.

One of the key assumptions of this special issue is therefore not so much that sexualized, racialized, and naturalized differences don't matter but rather that they no longer coincide with sexually, racially, and species-differentiated bodies. Advanced capitalism has de-linked the empirical referents of otherness (woman/native/earth other) from the imaginary institutions of sexuality/race/nature which traditionally framed them. It follows that the classical others are no longer the boundary-markers of categorical distinctions, or the necessary point of reference for the organization of a symbolic division of labor between the sexes, the races, and the species. Today, they have been transformed in the spectral economy of dematerialization of difference.

Genetic engineering and biotechnologies have seen to it that a qualitative conceptual dislocation has taken place in contemporary "Life" sciences. They have located the markers for the organization and distribution of differences in micro-instances like the cells of living organisms and the genetic codes of entire species. We have come a long way from the gross system that used to mark difference on the basis of visually verifiable anatomical differences between the empirical sexes, the races, and the species. We have moved from the biopower that Foucault saw exemplified in comparative anatomy to a society based on the governance of molecular *bios-zoe*-power of today. From disciplinary to control societies. In postmodernity, under the impact of the technological revolution, the political economy of the Panopticon is no longer adequate and has been replaced by the molecular informatics of domination. Regimes of discipline have metamorphosed into modulations of power that land on the living matter that composes the individual.[31]

This is not necessarily to say, however, that the patriarchal power relations have improved, or that the structural function which anatomical difference was called on to perform is obsolete. The collapse of traditional systems of marking difference makes it all the more urgent to reassert sexual difference as the privileged principle of alterity, of not-One as constitutive of the subject, and to elaborate nomadic forms of ethical accountability to match it. What is needed is an ethics of embodied differences that can sustain this challenge: an undifferentiated grammar of becoming simply will not do.

Difference, understood as the principle of not-One, fulfills the function of marking a fundamental break as the site of "origin" of the subject. What needs to be broken is the phantasy of unity, totality, and oneness. What is knocked out from the subject's psychic landscape is the delusion of stability, the phantasy of omnipotence. To recognize this basic, ego-deflating principle is ground zero of subject formation. The recognition of alterity in the sense of incommensurable loss and an unpayable outstanding debt to others entails the awareness that one is the effect of irrepressible flows of encounters, interactions, affectivity, and desire, which one is not in charge of. Bodily matter is always already sexed and hence sexually differentiated along multiple and heterogeneous axes.

This humbling experience of not-Oneness constitutive of the non-unitary subject, far from opening the doors to relativism, anchors the subject in an ethical bond to alterity, to the multiple and external others that are constitutive of that entity which, out of laziness and habit, we call the "self." The split or non-unitary nature of the subject entails the recognition of a prediscursive structure of the "self," of a necessary loss of that which is always already there – an affective, interactive entity endowed with intelligent flesh and an embodied mind. Whereas Lacanian psychoanalytic theory ontologizes this loss, and Derrida, with some help from Levinas, institutionalizes it as the constitutive interdependence of self and other, nomadic vital political theory builds on the productive aspects of the condition of not-One, that is to say a generative notion of complexity.

We have to learn to endure the principle of not-One at the level of the deep structures of our subjectivity. Becoming nomadic occurs by constructing communities where the notions of transience and passing are acknowledged in a sober, secular manner that binds us to the multiple "others" in a vital web of complex interrelations. Kinships systems and social bonding, like flexible citizenship, can be rethought differently and differentially, moving away from the fetishes of blood, earth, and origin implicit in the classical social contract. Given the extent of the transpositions brought about by advanced capitalism and the dislocations of traditional values and social bonding they have triggered, the conditions for a renegotiation of our being in *this* together are timely.

A materialist politics of difference-as-becoming today is a posthumanist, vitalist, non-unitarian, and yet accountable re-composition of *a missing people*. A community not bound together by the guilt of shared violence or by unpayable ontological debts, but rather by the compassionate acknowledgment of our shared need to negotiate processes of sustainable transformations with multiple others within the flow of monstrous energy of a "Life" that does not respond to our names, but endures through differences and by differing.

notes

Parts of this paper have been published in Rosi Braidotti, *Transpositions: On Nomadic Ethics* (Cambridge: Polity, 2006).

1 Paolo Virno, *A Grammar of the Multitude* (New York: Semiotext(e), 2004); Maurizio Lazzarato, "Immaterial Labor" in *Radical Thought in Italy: A Potential Politics*, eds. Michael Hardt and Paolo Virno (Minneapolis: U of Minnesota P, 1996).

2 Vandana Shiva, *Biopiracy: The Plunder of Nature and Knowledge* (Boston: South End, 1997).

3 Ulrich Beck, *World Risk Society* (Cambridge: Polity, 1999).

4 Anna Fausto-Sterling, *Sexing the Body: Gender Politics and the Construction of Sexuality* (New York: Basic, 2000) 235.

5 Gilles Deleuze and Félix Guattari, *A Thousand Plateaus: Capitalism and Schizophrenia*, trans. Brian Massumi (Minneapolis: U of Minnesota P, 1987).

6 Zillah Eisenstein, *Global Obscenities: Patriarchy, Capitalism and the Lure of Cyberfantasy* (New York: New York UP, 1998).

7 Martin Heidegger, "The Question Concerning Technology" in *Basic Writings*, ed. David Krell (New York: HarperCollins, 1993).

8 This term was coined by the avant-garde movement of the surrealists to refer to artworks and other forms of non-functionalist object production. The bachelor machine is sterile and pleasure-prone, not productive and re-productive.

9 Keith Ansell Pearson, *Viroid Life: Perspectives on Nietzsche and the Transhuman Condition* (New York: Routledge, 1997) 202.

10 Isabelle Stengers, *Power and Invention: Situating Science* (Minneapolis: U of Minnesota P, 1997).

11 Donna Haraway, *Simians, Cyborgs and Women: The Reinvention of Nature* (New York: Routledge, 1991) 61ff.

12 Rosi Braidotti, *Metamorphoses: Towards a Materialist Theory of Becoming* (Cambridge: Polity, 2002).

13 Nikolas Rose, "The Politics of Life Itself," *Theory, Culture and Society* 18.6 (2001): 1–30.

14 Mariam Fraser, Sarah Kember, and Celia Lury, eds, *Inventive Life: Approaches to the New Vitalism* (London: Sage, 2006).

15 Rosi Braidotti, *Transpositions: On Nomadic Ethics* (Cambridge: Polity, 2006).

16 Michael Hardt and Antonio Negri, *Empire* (Cambridge, MA: Harvard UP, 2000) 215.

17 For example, Genevieve Lloyd, *The Man of Reason: Male and Female in Western Philosophy* (London: Methuen, 1985).

18 See Deleuze and Guattari, *A Thousand Plateaus*.

19 Brian Massumi, "Requiem for our Prospective Dead (Toward a Participatory Critique of Capitalist Power)" in *Deleuze and Guattari: New Mappings in Politics, Philosophy and Culture*, eds. Eleanor Kaufman and Kevin Jon Heller (Minneapolis: U of Minnesota P, 1998) 60.

20 Donna Haraway, *Modest_Witness@Second_Millennium. FemaleMan©_Meets OncoMouse™: Feminism and Technoscience* (London: Routledge, 1997) 74.

21 Manuel DeLanda, *Intensive Science and Virtual Philosophy* (London: Continuum, 2002).

22 Elizabeth Grosz, *The Nick of Time: Politics, Evolution, and the Untimely* (Durham, NC: Duke UP, 1994).

23 Idem, *Space, Time, and Perversion: Essays on Bodies* (New York: Routledge).

24 This term is widely circulating in cyberspace to refer to the physical body, which is wet in relation to the dryness of electronic circuitry.

25 Judith Butler, *Gender Trouble: Feminism and the Subversion of Identity* (New York: Routledge, 1990).

26 More in Braidotti, *Transpositions*.

27 Luce Irigaray, *This Sex Which is Not One*, trans. Catherine Porter (Ithaca, NY: Cornell UP, 1985); *Ce sexe qui n'en est pas un* (Paris: Minuit, 1977).

28 Claire Colebrook, "Is Sexual Difference a Problem?" in *Deleuze and Feminist Theory*, eds. Ian Buchanan and Claire Colebrook (Edinburgh: Edinburgh UP, 2000).

29 Patricia MacCormack, *Cinesexuality* (Aldershot: Ashgate, 2008).

30 Braidotti, *Metamorphoses*.

31 Gilles Deleuze, "Postscript on the Societies of Control," *October* 59 (winter 1992): 3–7.

Index

Page numbers followed by 'n' refer to notes

absolute totality 4
Achilles: and tortoise 43
Agamben, G. 96
aggression 107–8
Amazonia 118–19, 129, 134
anachronism 36
androgyny 83, 121, 122, 125, 126, 128, 132, 137n
Ansell Pearson, K. 165
anthropocentrism 19, 20
anthropology 50, 129; philosophical 3
anticipation: non-dogmatic 142
Antigone 5, 23n, 74
antinomy: dynamic 4, 57, 58, 59, 65n, 128, 129; mathematic 58, 59, 65n, 127, 128, 129; sexual 3–6
anxiety 109–10
aphasia: hysterical 103; organic 103
appearances 57–8
asexual Being 55
attraction 86
authority: male 9
autonomous subject 48, 49
Avatar 11

baby-mother relationship 6
Badiou, A. 5, 6, 8, 98n, 141, 145–7, 154–6
barnacles 83–5, 90n, 91n; *Ibla* 84–5
Bataille, G. 7, 56, 119, 123
Beauvoir, S. de 11, 146
being 40, 45, 56, 57, 60, 92, 98n; commodification 61; in general 57
being or sex: and differences 48–66
Bell's palsy 102
bifurcation: sexual 83, 85, 86
Binswanger, L. 38
biocapital 99n
biocommunism 158
biologism 143
biology 3, 67, 71, 76, 86, 94, 95
biopolitical turn: and feminism 92–100
biopolitics 48, 93, 94, 95, 96, 99n; and sex 39, 59–62
biopower 17, 38, 39, 153, 169
biotechnologies 21, 169
bisexuality 132
Black Panthers 157
body 3, 44, 69, 109, 111, 119, 131, 155; and history of medicine 106–11; imaginary 107, 109; inventing 101–6; multinatural 123–4; psychoanalysis 101; reproductive capacities 79; sexed 61
bonus: incentive 45
bourgeoisie 15
Braidotti, R. 16, 27n, 164–71
Brown Blackwell, A. 77
Buddhism 147, 150
Butler, J. 16, 17, 51, 119

cannibalism 10
Cantor, G. 128, 129, 130, 132, 133
capital 22n, 61
capitalism 15, 38, 39, 60, 61, 141, 143, 154; advanced 164–5, 169; future 45; global 62; patriarchal 73
Casarino, C. 92–100
castration 11, 13, 20, 66n; complex 14, 37, 143
causality 4, 57
cerebral paralysis 101–2, 103, 104
Charcot, M. 105, 106, 107
Chinese dialectic: and sexuation 152–3
Chinese philosophy 147, 152
Christianity 142, 163n
citizenship 142, 158
civilization 115
classes 79
clinamen concept 64n
Colebrook, C. 168
commodification: being 61; sexed body 61
commodity: fetishism 96; life 96
communism 5, 15, 50, 143, 145, 158; Dao 155–7
Communist Party: Italian 145
communist universalism 141
compactness 43–5
comparativism: and intimation 132–3; plural and infinite 120

competition: male 80, 81
complexity 164, 165, 169
Confucianism 162n
Confucius 147–8
consciousness 35, 43, 53–4, 56
consumerism 146
consumption 49
continuity 43
convergent series 43
Copernican Revolution 2
Copjec, J. 3–6, 10, 17, 26n, 54–5, 57–8, 126, 152, 156; sexual compact 30–47
Corbin, H. 44
corporeal feminism 12
correlationalism 3
crisis: sex in 8–10
Cronin, H. 90n
cross-sex 121, 128
crustaceans 83
cultural life 94
cultural productions 73
cultural relations 87
cultural relativism 17
cultural studies 112
culture 8, 19, 59, 68, 71–7, 87, 97, 103, 111, 125, 133; of difference 153; moral codes 115; openness 78; Western 73

Dao 148, 149; of communism 155–7
Daoism 147, 148, 149, 150, 155, 157, 160n
Daoist vitalism 148
Darwin, C. 6, 7, 12, 75, 76, 79–88, 90n, 94
das Dasein 45
daughter-mother identification 14
Dawkins, R. 7, 89n
death 98; drive 51–4, 59, 97, 99n; personal 59
Debord, G. 61
decisionism 157
Deleuze, G. 18, 28–9n, 43, 59, 88n, 96, 118, 123, 134, 142, 157; and Guattari, F. 12, 20–1, 50–1, 52, 53, 126, 140, 165
Deleuzianism 48
democracy 69, 140, 142, 145
Derrida, J. 16, 18, 45, 115, 116
Descartes, R. 2, 41, 42
desire 19, 52, 53, 103–4, 108, 149
destiny: and woman 10–11
determinism: technological 34
Deutscher, P. 142
difference 143, 153, 164; gender 120; racial 12, 86, 153–5, *see also* sexual difference
dimorphism 86; sexual 7, 82, 83
dis-adapted evolution 6
discernment: female 81–2
disease 101, 106
disjunctive synthesis 125
dispositifs 34
diversity 118; phenotypic 12
divisibility 4
DNA 7
domination: informatics 165
drive 31–2, 97, 98; definition 41; theory 41
duality: sexual 151
dynamic antinomy 4, 57, 58, 59, 65n, 128, 129
dynamics 4

ecofeminism 154
ecology 158
ecosophy 155–6
egalitarian projects 71
egalitarianism 78, 79, 86, 90n
ego 42–3, 107, 108, 109, 110; formation 116n; ideal 107, 109
Ellis, H. 17
emancipation: women 10–11
embodiment: female 15–16
empiricism: transcendental 120, 129
enchanted materialism 166
Engels, F. 78
enjoying substance 41
Enlightenment 3, 13, 114, 140, 142; rationalism 8
envy: penis 9, 11
equality 70, 87
erogenous zone 43–5
Eros 52, 56, 59
essentialism 73, 103, 143
ethical intersubjectivity 16
ethics 3, 5, 6, 115, 158; pagan sexual 17
Euclid 3
eugenics 17
Eurocentrism 3, 146, 153
European patriarchy 22n
European Union (EU) 140
evil: radical 23n
evolution 7; dis-adapted 6; flight 9; woman 142
evolutionary psychology 77
Ex-Man 166
existentialism 73
exogamy 12
exploitation: women 72

family 74–5
fantasy 53
fascism 21
father: primordial 15
felicity conditions 119
Felman, S. 30
female discernment 81–2
female embodiment 15–16
female homosexuality 27n
female sexuality 9, 16, 39
feminine: virtual 169

femininity 4, 6, 9, 15–17, 19, 40, 54, 70, 127, 146, 150; Daoist conception 155
feminism 6, 11, 17, 21, 140, 144, 145, 146; and biopolitical turn 92–100; corporeal 12; fourth-wave 146; task 152
feminist revolution: goal 30
feminist theory: gender 30
feminist universalism 140–63
feminists: of difference 143; and psychoanalysis 30
fetishism 14; commodity 96
Fink, B. 65n, 139n, 152
Firestone, S. 30
flight: evolution 9
Foucault, M. 17, 18, 37–40, 92, 96, 101, 104, 106, 112–14
freedom 4, 5, 57
French Revolution 140, 144
Freud, S. 8, 13, 33, 35–6, 40, 43, 97, 102–3, 105–6, 109–10; French Freud 30
fundamentalism: Islamic 161n
fungi 8

Galileo 3
Gamble, E. 77
gametes 89n
gay rights 146
Geertz, C. 132
Geisteswissenschaften 103
gender 16, 46n, 103, 104, 110, 122, 129, 167, 168; construction 37; difference 120; feminist theory 30; identity 108; neutered 30–1; Papua New Guinea 118–39; theory 31, 34, 128
gender-sexuality 17
gender/queer theory 17, 18
genera 32
genetic engineering 169
genocide 154
German Idealism 142–3, 150
Gimi initiation ceremony 122–3
global capitalism 62
global risk society 164
globalization 164
God 4, 21, 32, 33, 46n, 95, 101
Goebbels, J. 154
Great Outdoors 3
Grosz, E. 3, 6, 12, 19, 49, 54, 67–91, 94–5, 97–8, 143
growth 122, 135n
Guattari, F. 50, 141, 155–6, 157; and Deleuze, G. 12, 20, 21, 50–1, 52, 53, 126, 140, 165

Haeckel, E. 9
Hallward, P. 21, 133–4, 145
Haraway, D. 165, 166, 167
Hegel, G.W.F. 18, 50, 73–4, 75, 144
Hegelianism 87
Heidegger, M. 45, 54, 68, 92, 165
Heraclitus 34
hermaphroditism 8, 25n, 83, 84, 85, 86
hermeneutics 49
Hinduism 147
historicism 17
Hitler, A. 154
homology 130; vs recursivity 131–2
homophobia 21
homosexuality 17; female 27n
human nature 33
humanism 22n, 50
humanity function 156
humans 92, 98, 166
hysteria 10, 102–3, 104, 105–6, 106, 109; symptom 107, 109, 110, 116
hysterical aphasia 103
hysterical paralysis 103, 105, 106, 109

id 43, 51
ideal ego 107, 109
idealism: German 142–3, 150
identification: secondary 107, 108; with signifiers 108
identity 112; gender 108
imaginary body 107, 109
impossible 126–8
impotence 13
incentive bonus 45
incest 9, 11, 54; definition 126; prohibition 12, 59, 60, 66n, 126, 133; taboo 12
India 147
individuation 33
Industrial Revolution 153
infinity 129, 138n; intimated 129–31; Iqwaye 119, 132
informatics of domination 165
initiation ceremony: Gimi 122–3
intellectual perception 42
interpersonal relations 133
intersubjectivity: ethical 16
intimated infinity 129–31
intimation: and comparativism 132–3
Iqwaye: counting system 129–30; infinity 119, 132
Irigaray, L. 2–3, 15–16, 68–77, 86–8, 89n, 92, 140–2, 145–7, 150, 156
Islam: fundamentalism 161n; philosophers 32–3
Islamophobia 146, 161n
Italian Communist Party 145

jouissance 26n, 27n, 33, 37–8, 41, 42, 43, 44, 55
justice 115

Kant, I. 2, 3, 4, 18, 54, 55, 57, 58, 127
Karatani, K. 60
Kazarian, E. 62n

Kiarina Kordela, A. 48–66
Kierkegaard, S. 34
kinship: practices 11–12; systems 170
Kirsner, D. 101
Klein, M. 10, 13
Kojève, A. 52
Kristeva, J. 157

labor: sexual division 8
labor-power 60, 61
Lacan, J. 12–15, 33, 38, 41, 43, 58–9, 104, 108, 114–15, 123–4
Lacanian theory 49
Lacanianism 48
lack 51–4
Laclau, E. 144
language 92, 93, 104, 107, 111, 115
Laozi 148
Laplanche, J.: and Pontalis, J-B. 50, 51, 53
latency period 35
Lauretis, T. de 30
law 6, 140; function 52; and sexuality 60, 112–16; social 74; and transgression 114
legalism 162n
Leibniz, G.W. 44
lesbian phallus 28n
lesbianism 11, 19
Lévi-Strauss, C. 6, 11–12, 54, 55, 59, 125, 126
liberation 76; sexual 37, 38
libidinal energy 13
libidinal normalization 107–8, 109
libido 37, 43–4, 45, 55, 59; narcissistic 108; role 109; sexual 108; theory 9
life 92, 93, 94, 96; as commodity 96; natural 94
limit 43, 44, 127
linguistic turn 92, 93, 96, 98n
love 44–5, 141, 156

MacCormack, P. 168
male authority 9
male competition 80, 81
male sex 65n
maleness 132
mammals 83, 90n; sexual difference 9
man 69, 88n; as camera obscura 2–3
Maoism 145, 146
Margulis, L. 7
Marx, K. 20, 32–3, 60, 78–9
Marxism 15, 73, 87, 145
masculinity 10, 14, 17, 54, 70, 127, 150
masochism 11, 19
Massumi, B. 149, 166
materialism 164; enchanted 166; vitalistic 167
mathematic antinomy 58, 59, 65n, 127, 128, 129
mathematics 131, 138n
matriarchy 11
matter-realism: and vital politics 166–7
Mayr, E. 7
Mead, M. 12
media 21
medicine: history and body 106–11
Meillassoux, Q. 3
Mein Kampf (Hitler) 154
Melanesia 121–2, 123, 129, 134
melodrama 10
memory: phylogenetic 9; transcendental 123
memory system 35
menstruation 11
Mill, J.S. 78, 88
Mimica, J. 129–32
misogyny 5
Mitchell, J. 37
mitochondria: DNA 7
mitochondrial Eve 7
Möbius strip model 94, 99n
modernism 55
monotheism 32
monotheistic sex 37, 46n
moral codes 115
morality 5, 8
mortality 58
mother–daughter identification 14
motor paralysis 101
multiculturalism 144
multiculturalist pluralism 129
multiculturalist uninaturalism 118, 119
multinatural body 123–4
multinaturalism 119–20
multiplicity 21
mushrooms: *Schizophyllum commune* 8
myths 54

n sexes 20–1
narcissism 51–4
narcissistic libido 108
nation-state 6
nationalism 146, 161n
natural life 94
natural selection 6, 7, 76–80, 82, 85, 97, 98
naturalism 73, 143
nature 13, 19, 59, 60, 97, 103, 125, 133, 147; dualization 150; sexual difference 67–91
Naturwissenschaften 103
Nebenmensch 42, 44
Neo-Daoism 151
Neo-Spinozists 49
nervous system 101, 102
neurology 103
neurosis 8, 104
neutered gender 30–1
New Ageism 11
Newton, I. 3
nomadism 168
nominalist position 32

non-dogmatic anticipation 142
nonhuman sex 20
nuclear family 8

objects 121–2, 123, 129, 131
Oedipalization 10, 12, 50–1
Oedipus: complex 9, 25n, 108; myth 54, 55
Oedipus the King (Sophocles) 9
Omalyce 132
ontogeny 9
ontology: sexual difference 70
openness 78
oppression 76, 92
organic aphasia 103
organic paralysis 105, 109
organism 109, 111
organs: sex 7
orgasm 25n
Other 19, 20
otherness: sexed 18–20
over-determination 31

pagan sexual ethics 17
pansexualism 31
Papua New Guinea: gender 118–39
paralysis: cerebral 101–2, 103, 104; hysterical 103, 105, 106, 109; motor 101; organic 105, 109; periphero-spinal 101–2, 105; representation 102, 103
patriarchal capitalism 62
patriarchal power relations 169
patriarchy 10, 16, 69, 70, 74, 77, 78, 95, 141, 143, 145, 146; European 22n
penis 122; envy 9, 11
perception 35; intellectual 42
performativism 119
periphero-spinal paralysis 101–2, 105
perspective 126–7
perversion 8
phallic stage 13
phallocentrism 9, 13
phallus 16, 20, 27n, 29n, 123, 143; lesbian 28n; universality 14
phenomenology 18, 19, 73, 74
phenotypic diversity 12
philosophers: Islamic 32–3
philosophical anthropology 3
philosophy 3, 86, 95, 111
phobias 34–5
phylogenetic memory 9
phylogeny 9
place 5–6
pluralism: multiculturalist 129
political power 148
politics 3, 141, 142, 158; vital and matter-realism 166–7; *zoe*-posthuman 166
polymorphism 86
Pontalis, J-B.: and Laplanche, J. 50, 51
population thinking 7
pornography 21
postmodern subject 48, 49
postmodernism 55
postmodernity 165
poststructuralism 141
power 28n, 38, 153; political 148; sex 15–18; woman 156
power relations 164; patriarchal 169
primordial father 15
prohibition: incest 12, 59, 60, 66n, 126, 133; self-referentiality 60
proletariat 15, 158
psychic functioning 36
psychoanalysis: and vitalism 48–51, 119
psychology: evolutionary 77
psychotherapy 17

queer theory 17, 18, 168

race 22n, 79, 141, 154, 169
racial difference 12, 86, 153–5
racial tropes 142
racialization 2
racism 161n; latent 9
radical evil 23n
rape 6, 8, 81, 158
rationalism 8
real universal 33
reason 2, 42
recapitulation 25n
reciprocity: sexual 145
recollection 34
recursivity: vs homology 131–2
regress: *in infinitum* 58
regression in indefinitum 58
relation 120–2, 134
relationality 18
relativism: cultural 17
relativity theory 3
religion 56, 61, 62, 111, 140
repetition 34, 35, 36, 37, 98, 167; denial 34
representation 112, 134; paralysis 102, 103
repression 38, 114; hypothesis 37, 60
reproduction 84; body capacities 79; sexed 59; sexual 7, 24n, 55
res cogitans 42
rhizomatics 49
rights 142, 144, 146, 153, 158; gay 146; women 144, 146, 153
risk society: global 164
Riviere, J. 10
Rolland, R. 42
Russian Revolution 51

Sacher-Masoch, L. 19

Sade, Marquis de 3
sadism 10, 19
Saldanha, A. 1–29, 140–63
same-sex 128, 168
Sartre, J-P. 53, 61
schizoanalysis 20, 21
Schizophyllum commune 8
Schor, N. 140
scientia sexualis 38, 39
Scott, J. 144
secondary identification 107, 108
selection: natural 6, 7, 76–80, 82, 85, 97, 98; sexual 6–8, 67, 68, 86–8, 94, 95, 97, 98
self-referentiality 129; prohibition 60
set theory 59, 64n, 128–9
sex 92, 103; or being and differences 48–66; and biopolitics 39, 59–62; in crisis 8–10; monotheistic 37, 46n; organs 7; and phenotypic diversity 12; power 15–18; and sexual difference 55–9; theory 31, 33
sexed body: commodification 61
sexed otherness 18–20
sexed reproduction 59
sexes 79
sexism 14, 21
sexual antinomies 3–6
sexual bifurcation 83, 85, 86
sexual compact 30–47
sexual difference: asymmetrical 15; logical dimension 4; now 21–2, 54–5; philosophical concept 68–9; principle of not-one 167–70; theory 16; two 45
sexual dimorphism 7, 82, 83
sexual duality 151
sexual equality 70
sexual *jouissance* 56
sexual liberation 37, 38
sexual practices 31
sexual reciprocity 145
sexual relations 42, 86, 95, 123–4; impossibility 118
sexual reproduction 7, 24n, 55
sexual selection 6–8, 67, 68, 76–88, 94, 95, 97, 98; origin 83–6
sexuality 11, 20, 21, 101, 104, 107, 108, 110, 125, 167, 168; di-phasic onset 35; female 9, 16, 39; feminine 45; forms 83; human 8; infantile 17; and law 60, 112–16; normative form 113; onset 36; Western 6
sexualization 2
sexuation 14, 15, 16, 43, 54, 156; and Chinese dialectic 152–3; formulas 126–7
Shakespeare, W. 116n, 117n
Shepherdson, C. 17–18, 101–17
signifiers: identification 108
Sinoza, B. 97
slavery 78, 154
social constructionism 17
social contract theory 115
social law 74
social order 68, 72, 74, 75
social organization 76–7, 145
social relations 71, 87, 143
sociality 119
society of the spectacle 61
sociobiology 77, 78
socius 125–6
Song, H. 118–39
Sophocles 9, 54
soul 118
sovereignty: subjective 2
space 3, 4, 57, 58
species 32, 80, 81; relationship 118, 119
speech 86, 133
Spinoza, B. 49, 95
spiritualization 74
Spivak, G. 143
Stone, A. 150
Strathern, M. 118, 120–3, 125, 126, 128, 129, 131, 133
structuralism 12, 50, 73
subjective sovereignty 2
subjectivity 3, 14, 49; masculine 15
sublation 57
subversion 143–5; universality as 143–5
superego 115
supernumerary One 125

taboo: incest 12
Taoism 162n
Tarzan 1–2, 8, 12, 20, 22
Tarzan of the Apes 1
technobodies 167
technological determinism 34
technologies 165, 167; of gender 34
territorialities 52
theosophy 154
thermodynamics: second law 7
thinking: population 7; typological 7
thinking substance: function 42
third substance: myth 40–3
time 3, 4, 35, 36, 57, 58, 127, 128; third synthesis 51, 52
tissue theory 106
transcendence 96
transcendental empiricism 120, 129
transcendental memory 123
transcendental turn 3
transgression 52, 101, 104, 112–13; and law 114
transsexuality 20
transvestitism 14
tropes: racial 142
truth: of psychoanalysis 111–12
typological thinking 7

unconscious 8, 17, 42, 43, 49, 50, 53, 56, 105, 110
under-determination 31
uninaturalism 133; multiculturalist 118, 119
universalism 140, 143, 144, 146, 153, 154, 155; communist 141; feminist 140–63; white 153–4
universality: phallus 14; as subversion 143–5
universalization 145

Virno, P. 60, 61
virtual feminine 169
vitalism: and psychoanalysis 48–51, 119
vitalistic materialism 167
Viveiros de Castro, E. 118, 119, 123, 126, 127, 131, 133

Wagner, R. 129, 131
Wallace, A.R. 83
Wang Bi 150, 151
Warring States era 148
wealth 73
Western culture 73
Western sexuality 6
Westoxification 46n
White Man 140, 155
white universalism 153–4
Whitford, M. 142
Wittgenstein, L. 93
Wittig, M. 96
Wollstonecraft, M. 144
woman 19, 28n, 40, 69–70, 75, 88n, 142; becoming 157, 169; and destiny 10–11; emancipation 10–11; evolution 142; exploitation 72; male treatment 78; power 156; rights 144, 146, 153; sexual economy 168
womanliness 10
womb 122

xenophobia 6

yin-yang 140, 142, 149–52

Zeno 43
Žižek, S. 4, 15, 21, 28n, 49, 57, 143, 150–2, 154, 155–7, 163n
zoe-posthuman politics 166

For Product Safety Concerns and Information please contact our EU representative GPSR@taylorandfrancis.com Taylor & Francis Verlag GmbH, Kaufingerstraße 24, 80331 München, Germany

Batch number: 10726412

Printed by Printforce, the Netherlands